Questions About Life and God

Bryant Harris

Published by Kinetic Digital Publishers

www.kineticdigitalpublishers.com

For permissions, inquiries, or other correspondence, please visit our website.

TABLE OF CONTENTS

Introduction 1
1. Is there a God? 2
2. Is the Bible real and if so what are the scientific reasons for the Bible? ... 9
3. A What proof is there Jesus Christ from a biblical perspective? 20
3. B What is proof of Jesus Christ from a non-biblical perspective? 31
4. Why did God put Jesus earlier in this world? 37
5. Why did Jesus come once, and why did He come to Israel? 41
6. Why is there suffering? 46
7. Why do innocent people die? 50
8. I understand why sinners don't die, but what about murderers? Why do they live? 54
9. Why don't we see miracles, or why don't I see the wonders of God? ... 57
10. Why don't my prayers get answered? 59
11. What is the trinity? 62
12. Why do we pray? 65
13. How to pray? 70
14. How do you know you are a good Christian? 73
15. What can you do to become a good Christian? 80
16. Why are marriages failing? 83
17. Is there a hell, and why does God allow hell? 93
18. Why does Satan try to kill, steal, and destroy? Why does he not like humans 98
19. What is the state of the world? Is the world is getting worse? 102
20. Are we in the end times? Is the world ending? 115
Final Thoughts 125

INTRODUCTION

I wanted to write this book for several reasons. I wanted a book for people who struggle with their spirituality. I also wanted to answer the big questions that life has. No matter where you are in your walk with God, this will help you in your journey. What I have done is ask questions to people who don't have faith or are struggling with faith. I answer these questions based, on scientific reasoning, and sometimes on logic. I don't pretend to know everything. I am not God. Sometimes you may disagree, and that is ok. There are twenty questions that I will answer.

It will start with why I believe there is God. Then it goes to the scientific proof of the Bible. If you can't believe in God, then you certainly are not going to believe in the Bible. The questions are designed to build your faith question by question. Then we also tackle questions like how to pray, and we also answer questions that people of faith struggle with. I believe when we die, we will find out at least some of these questions.

That is my prayer that you will find salvation through our LORD JESUS Christ. I pray that this book helps you.

1. IS THERE A GOD?

The Big Bang theory was a theory that was proposed to explain the beginning of life. It was an atheist way of explaining the way life began. If we could disprove the Big Bang theory, then we could prove God. We will also argue that atheists and agnostics believe in God. Atheists also argue that we all evolved from single-cell organisms. We will see what the problems with evolution are.

If you believe in the Big Bang. Then fundamentally, you believe you could create something out of nothing. In science, matter could not be created or destroyed. Right now, the only person who could do that is a God. Science could not explain something out of nothing. Only God can be so fundamental that everyone should believe in God or a higher being. The question is what type of God do you believe in? Do you believe In God that lives, or do you think God is dead? In that way, atheists believe in God that is dead.

[1.] The Big Bang theory states that matter all started with an atom, and there was an explosion. From that explosion, life began from that one atom. Scientists base this upon seeing the universe expand. They see that the edge of the universe is moving faster than galaxies that are not on the edge. They also based this upon leftover heat from what they propose. I would say most people don't know that most of the science disproves this.

Have you ever seen an explosion before? When you see an explosion, does it just go in one direction, or does it go in all different directions? If there was an explosion that created life, then you would expect the center of the universe to be the earliest part of the universe and the universe to be in a circle. [2] Reality is that the universe, from the earliest to oldest, is left to right. Or right to left, depending on your orientation. If God was making the universe, it would make sense that the universe would be older from left to right.

If we believe in the Big Bang. Conditions had to be just right. [3] The beginning force that would explode had a 10^{59} chance the universe would have collapsed on itself, If the gravity was 10^{33} weaker, then none of the stars would be produced. If the nuclear strength in relation to gravity was weaker, most of the hydrogen would not be turned into helium, which would probably mean no water. If the nuclear force were just 2 percent stronger, it would have prevented the formation of protons. If it had been 5 percent weaker, we would have no universe without stars.

Compare that to people who had a religious experience. [4]In 2006, Ipsos did a survey where they asked people who had a religious experience to 47%. In the Pew Forum, it was 49%. So it would take a lot of faith to be an atheist. There is less than a 10^{59} chance that the Big Bang happened. About 1 out of 2 people had a religious experience that science could not explain. Let us explore evolution and see the flaws in it.

Have you ever had a bath? When you have a bath, it is just hot on one side and cold on the other side, assuming the water faucet is shut off. That is convection that you would expect to be evenly warm throughout. We don't see that in the universe. We see cosmic radiation, which is not dispersed evenly. If we believe the Big Bang, then this cosmic radiation should be even. You should also see an even distribution of stars and planets, but you don't see that.

Evolution could be argued to have started with Charles Darwin's Origin of Species. If science needs to answer the question of where life begins, that is central to that.

Contrary to what you might have believed, there is no consensus that all life was created from one being. There has been much debate from the beginning when Charles Darwin came out with "The Origin of Species."

One of the biggest criticisms of Evolution is the beginning of life. According to Evolutionists, they think that life began when chemicals mixed together. They feel like when lightning hits, change it, and that is how life was created. [5]The problem is when they tried that in the experiment by Stanley Miller and Nobelist Harold Urey (1953) tried to have experiments to create amino acids, and they were only able to get some of the amino acids. There have been replications since then he simplest life form has 437 DNA parts. Scientists can only replicate 32% of the DNA in a natural experiment. Science can make life out of nothing, but in a natural way.

In 1802, William Paley's "Natural Theology." William Paley's argument was that if you saw a watch lying on the ground, what would you think? The watch has many parts and is bright gold. Would you think nature creates that gold watch, or would you think a craftsman makes it?

Going further, we have an estimated anywhere between 2 million and 50 million land species that lived on this earth. There are about estimates of over 400,000 plant species on the earth. [6] Animals range from 2 chromosomes (roundworm) to some butterflies that have about 450 chromosomes. You may be surprised that plants have chromosomes as well; they range from the sunflower, which has 4, to Mulberry that has about 308 chromosomes. Science has yet to find an organism that has changed the number of chromosomes. That is a serious flaw for the evolutionists.

Scientists have yet to find an organism that went from increasing their chromosomes. In order to believe evolutionists, you would need that. On this earth, we ranged from 4 to 308. In order to believe in evolution. This would have to be a common experience.

Scientists have yet to find the reason why organisms evolved to have feathers, or the first organism to have a requirement for life, or the first organism to fly. If you believe evolutionists, there must be millions of first creatures. In reality, you don't see that.

Evolutionists would have you believe that all scientists believe in evolution. The thing is, a lot of scientists don't believe in evolution. Evolution is not a law, it is just a theory.

Webster's defines science as *knowledge about the natural world that is based on facts learned through experiments and observation.* When we date things, we can not know for sure that it is accurate. There was not anyone that was here 4.5 billion years ago. When we age things that are really old, we use the uranium half-life. We make assumptions that uranium has decayed the same for this period. According to the Bible, the earth was here before he got started making things in 6 days. **Genesis** 1:2. "And the earth was without form, and **void**; and darkness was upon the face of the deep. And the Spirit of God moved upon the face of the waters.

When they estimated the dinosaur's life, they estimated it based upon Uranium. What if I told you they had found dinosaurs with skin on their bones? In 2011, in Canada, they found a Nodosaur dinosaur. This dinosaur had its skin on as well as its scales.

What if I told you that they have found dinosaurs in ancient cultures? In a Buddhist temple in the 12 th century. There are figures of what appears to be a stegosaurus. This figure is carved into the temple. You also have figurines in Ancient Mexico 2500 BC carbon dated. You also see these pictures of dinosaurs on ancient stones in Peru. You also see it in the Anasazi people in Utah. They have a picture of a dinosaur etched into the rock.

In the Bible, it appears that Job was describing a dinosaur

15 Behold now behemoth, which I made with thee; he eateth grass as an ox. 16 Lo now, his strength is in his loins, and his force is in the navel of his belly. 17 He moveth his tail like a cedar: the sinews of his stones are wrapped together. 18 His bones are as strong pieces of brass; his bones are like bars of iron.

You also hear a lot of cultures talk about dinosaurs in their language

Name	Description	Location or Language
Aziwugum	Giant reptile	Innuit
Bax'an	Terrible water monster	Dakota Sioux
Behemoth	Giant swamp reptile	Hebrew (Job)
Drakon	Dragon	Greece
Grendel	Swamp monster	Denmark
Knucker	Swamp dragon	Wales
Long	Dragon	China
P'ih mw	Giant reptile	Egyptian hieroglyphs
Ro-qua-ho	Giant reptile	Iroquois
Smok	Dragon	Poland
Uk'tena	Homed water monster	Cherokee[2]
Worm (voorm)	Dragon	Germany

Why would all these cultures have words that described dinosaurs unless they really had an animal like that? When they found dinosaurs that were more intact, they found animals.

As we said, scientific knowledge of the natural world is based on facts learned through experiments and observation. We could observe that dinosaurs really did exist along with humans. We could observe that dinosaurs really did exist more recently than what scientists are saying.

[7]In a massive genetic study, senior research associate at the Program for the Human Environment at Rockefeller University, Mark Stoeckle, and University of Basel geneticist David Thaler took 5 million samples of 100,000 animal samples. They found that about 90% of all animals came into existence about the same time, 90,000 to 100,00 years ago. Again, the scientists are using a half-life for a substance that can't be observed.

Evolutionists would argue that humans have been around for thousands of years. DNA changed around the time when civilization started popping up. Once civilization started popping up. Sumerian was the first civilization; they invented the wheel, writing, geometry, irrigation for farming, boats, leather, saws, axes, knives, swords, and boots. Clearly, you could see that these humans were different. Then those so called humans before. The DNA shows that they were different.

People don't realize how powerful we are. Imagine if you could fit the world's knowledge into a massive computer the size of two square blocks. That same amount of world knowledge can fit into a human brain. The human brain is about 20 watts of electricity, vs the best computer, which holds a lot of electricity. [8]There are an estimated 50,000 - 70,000 thoughts that humans have. We have 86 billion neurons in our brain. (86 billion is like going over the moon 179,992 times) Nerve signals can go faster than 150 miles per hour. If DNA was uncoiled, it would stretch from 10 billion miles away from Earth to Pluto and back.

A lot of people don't think of the Bible literally when it says life began in less than a week. I have given you evidence that points to that dinosaurs roaming the earth with us. We are complicated beings that are unique from anything else created in the universe.

What about the rest of the week? What evidence is there about the solar system? That shows the universe was created in days. They estimate there are 10^{21} stars. Yet we have not observed one. They think stars form from dust clouds, yet they have not seen one form. Now we have observed stars die. If science is correct about the universe, then we should see many more dead stars, but we don't. We also would expect that with $10^{21,}$ we would have seen one star created in all of that.

Based on the Big Bang theory and how science explains stars being formed. You would not see binary stars. Scientists say that stars are formed based on dust clouds. If that is the case, then there should not be any binary stars. Some estimates suggest that 85% of all stars are red dwarfs, and that of the red dwarfs, 25% are binary systems.

Conclusion

For us to believe there is no God. We have to believe in 1 $10.^{59}$ chance when the Big Bang happened that the universe would not collapse on itself. On top of that, if there 10^{33} weaker than none of the stars would be created. Or if it was two percent stronger than the formation of the protons.

Going further, in order for there to be life, the conditions had to be just right. Evolutionists still can't create the first forms of life. The Bible doesn't contradict that animals have not adapted over time. But there is a huge difference between adaptation and evolution. There are between 2 and 50 million different animal species and 400,000 different plant species.

To date, you don't see the first missing links in the fossil record. 90% of animals just appeared out of nowhere. You don't see the first animals that have eyes or wings or feathers.

Science is the study of what you could observable see. You could observe that the animals they said died out millions of years ago were still present here with humans. From all cultures, they have reported some kind of dinosaur like creature, including pictures and figurines.

. 17 He moveth his tail like a cedar: the sinews of his stones are wrapped together. 18 His bones are as strong pieces of brass; his bones are like bars of iron15 Behold now behemoth, which I made with thee; he eateth grass as an ox.

There is something really different between humans and animals. We have over 86 million neurons in the brain. If our DNA were uncoiled, it would stretch to Pluto and back.

Also, fundamentally, we must all believe in God or a higher being because there is no explanation for making something out of nothing except there being a God. We may debate what kind of God there is, but I think there is enough evidence to support the existence of God.

2. IS THE BIBLE REAL AND IF SO WHAT ARE THE SCIENTIFIC REASONS FOR THE BIBLE?

You may feel that there is a God. But you may be skeptical when it comes to the Christian God. There is no mistake that we started off with many religions through the years, and the Christian religion ended up. Like a trial-and-error process, people have moved from a poly- God system to a monotheistic religion.

Sceptics of the Bible have stated that the Bible is full of "old-fashioned thinking." What if I told you that the Bible told about science 100 years and in some cases thousands of years, before it was proven scientifically

Leviticus 13 4-8 Message **"If the shiny spot on the skin is white but appears to be only on the surface and the hair has not turned white, the priest will quarantine the person for seven days. On the seventh day, the priest will examine it again; if, in his judgment, the sore is the same and has not spread, the priest will keep him in quarantine for another seven days. On the seventh day, the priest will examine him a second time; if the sore has faded and hasn't spread, the priest will declare him clean—it is a harmless rash. The person can go home and wash his clothes; he is clean. But if the sore spreads after he has shown himself to the priest and been declared clean, he must come back again to the priest, who will conduct another examination. If the sore has spread, the priest will pronounce him unclean—it is a serious skin disease and infectious.** It talked about quarantining people and staying away from sick people. They started quarantining people in the middle ages. When ships would come in and bring disease. That Bible verse also talks about going home and washing clothes.

Leviticus 15-13.NLV 13 **"When the man with the discharge is healed, he must count off seven days for the period of purification. Then he must wash his clothes and bathe himself in freshwater, and he will be ceremonially clean.** When they say fresh water, they are referring to running water. Using fresh running water wasn't discovered until 1845. Where Dr. Ignaz found that women died more in childbirth when the doctors did not wash their hands. When the doctors washed the hands, they found the woman was less likely to die. Leviticus was said to be written around 1400 BC. Leviticus also specified running water, not just still water.

*Leviticus 1 -3 (New Living translation)***1 The LORD said to Moses, 2"Give the following instructions to the people of Israel. If a woman becomes pregnant and gives birth to a son, she will be ceremonially unclean for seven days, just as she is unclean during her menstrual period. 3 On the eighth day, the boy's foreskin must be circumcised.** Also *Luke 2: 21(NLV)* **Eight days later, when the baby was circumcised, he was named Jesus, the name given him by the angel even before he was conceived**. It talks about circumcision. The Mayo Clinic says that the health benefits of circumcision

- Easier to clean a circumcised penis vs a penis that is not
- Decrease risk of urinary tract infections
- Decreased risk of sexually transmitted diseases

Also, in the Bible, it says wait till eight days. Doctors say that is when babies are able to form blood clots. This professor, H. Dam, discovered vitamin K and blood clotting. Blood clotting has been tested medically to start around day eight of the baby's life.

Leviticus 12-14 **"You must have a designated area outside the camp where you can go to relieve yourself. 13 Each of you must have a spade as part of your equipment. Whenever you relieve yourself, dig a hole with the spade and cover the excrement. 14 The camp must be holy, for the Lord your God moves around in your camp to protect you and to defeat your enemies. He must not see any shameful thing among you, or he will turn away from you.**

That passage in Leviticus talks about having your excrement out in the field. There are estimates of over 250,000 people who die each year, which could be eliminated by the simple elimination of sewage near someone's house. It is said that cholera, diarrhea, dysentery, hepatitis A, typhoid, and polio are most commonly associated with proper sanitation.

In the world, they thought it was held up by a god name atlas. But the Bible talks about the Bible *JOB 26 -7* "**the Earth hangs on nothing** " which is scientifically correct. Also in *Isaiah 40-22* **God sits above the circle of the earth.** There were many Ancient societies, though the earth was flat. Things changed when Christopher Columbus sailed to North America in 1492.

Believe it or not, but the Bible may be talking about Dinosaur. The Bible did not use the word dinosaur however, it described one In *Job 40 15-24* **"Take a look at Behemoth, I made, just as I made you. It eats grass like an ox. Its powerful loins and the muscles of its belly. The tail is as strong as a cedar. The sinews of its thighs are knit tightly together. Its bones are tubes of bronze. Its limbs are bars of iron. It is a prime example of God's handiwork, and only its Creator can threaten it. The mountains offer their best food, where all the wild animals play. It lies under the lotus plants, hidden by the reeds in the marsh. The lotus plants give it shade among the willows beside the stream. It is not disturbed by the raging river, not concerned when the swelling Jordan rushes around it. No one can catch it off guard or put a ring in its nose and lead it away.**

Some people would say that it would be a hippopotamus. It was believed to be a hippopotamus at the time because that was the largest animal known to be around the Jordan River. But the scripture says the tail was like a cedar(tree). The scriptures, it talks about a massive beast. Something that would be an herbivore eats plants. Something that would drink massive amounts of water. Something that would have been able to wade through the Jordan River. From the Bible, it sounds like Brontosaurus would be the best example of the animal explained. Bones for the Brontosaurus weren't found until 1879. This was so many years ago, and it would be hard to use words to describe things. Verse 19 says "**that he that made him can make his sword to approach unto him.**" So God has the ability to kill this behemoth. Reading between the lines God killed the dinosaurs or allowed them to die.

In the Bible *Leviticus 17 14* **For it is the life of all flesh; the blood of it is for the life thereof: therefore I said unto the children of Israel, Ye shall eat the blood of no manner of flesh: for the life of all flesh is the blood thereof: whosoever eateth it shall be cut off.** A couple things from that statement state how important blood is. It was a practice a couple of hundred years ago to remove blood. Bloodletting was used by the ancient Greeks used, it was used in Medieval times as well. On December 13 1799, George Washington, our first president, had a sore throat. To fix the sore throat, they started taking blood from him. It is estimated that they took 5-7 pints of blood in 16 hours. He died 3 days later. It most likely died because of the bloodletting technique.

Leviticus 17-15 **Every person who eats an animal which dies [of natural causes] or was torn by a predator, whether he is native-born or a stranger, he shall wash his clothes and bathe in water, and be [ceremonially] unclean until evening; then he will become clean.** This verse talks about cleaning yourself when you eat an animal. We know from science how important it is to clean your clothes. Because blood carries many diseases.

Leviticus talked about Leprosy. Leprosy was skin disease. Depending on the severity of the disease. *(Leviticus 13-4)* **4 If the bright spot is white on the skin of his body and does not appear to be deeper than the skin, and the hair on it has not turned white, the priest shall [b]isolate the person who has the infection for seven days. 5 The priest shall examine it on the seventh day, and if in his estimation the infection has not changed and has not spread on the skin, then the priest shall isolate him for seven more days. 6 The priest shall examine him again on the seventh day, and if the infection has a**

more normal color and the spot has not spread on the skin, the priest shall pronounce him clean; it is only a scab; and he shall wash his clothes and be clean.

The Black Death was a horrible death, killing 75-200 million people in the Middle Ages. It is estimated that half of the Europeans died because of this. We know that if the quarantine were not implemented, millions of lives wouldn't be saved. It wasn't until about 1892 that the US government began quarantining people.

What if I said the Bible talked about our modern world? In *Revelation 11 9-10* **9 And for three and a half days, all peoples, tribes, languages, and nations will stare at their bodies. No one will be allowed to bury them**. I think there are some important lines there. All the tribes' languages and nations stare at the body. How can that be without having the invention of TVs.

What about other religious books? Other religious books don't have things that have been proven scientifically, like the Bible has. In later chapters, we will discuss even more how the Bible tells the future. Many religions force the user to go through the course of life. The Bible and the holy spirit can guide your life. Most of the religions, the people who founded the religion are dead. The holy spirit is still alive.

It is also no surprise that the Bible is the number one book sold. In 2005, the Bible had more than 25 million copies. That more than doubled the sales of the second-best-selling book, Harry Potter. The Bible hasn't changed, it has been the same. The Bible has been translated into over 500 languages.

Let's examine Islam. Islam has a lot of the same prophecies that we talked about earlier. In the Quran mentions Jesus around 25 times. It mentions the Prophet Muhammad 4 times. In the Koran, Jesus was born by a virgin birth. **And [make him] a messenger to the Children of Israel, [who will say], 'Indeed I have come to you with a sign from your Lord in that I design for you from clay [that which is] like the form of a bird,**

then I breathe into it, and it becomes a bird by permission of Allah. And I cure the blind and the leper, and I give life to the dead - by permission of Allah. And I inform you of what you eat and what you store in your houses. Indeed, that is a sign for you, if you are believers. *Quran 3 49.*

This confirms a couple of things. In this verse, Jesus is called a messenger of Israel. So even with Islamic Jesus was a prophet. According to Islam and Christianity, Jesus was able to raise from the dead. Islam also provides proof that he was able to heal the sick. Now, according to Islam, Muhammad was more better profit than Jesus. According to the Quran, Muhammad performed many miracles, but none involved giving life. One of the miracles attributed to Muhammad is splitting the moon in half. There is no scientific basis for him splitting the moon. I personally think that if the moon were split and put back together, there would be some scientific evidence.

What about the Bible? Surely there would have to be some fantastic things that could be unproven, For example, in the Bible, it talks about unicorns. **"God brought them out of Egypt; he hath as it were the strength of an unicorn."** *Numbers 23:22* **"But my horn shall thou exalt like the horn of an unicorn: I shall be anointed with fresh oil."** *Psalm 22:10.* There are more passages than that. The definition of unicorn has changed over time. If you were to look up the definition of unicorn, they would say a one-horned mythical creature. That definition has changed over time, *Webster defined* unicorn as 1. an animal with one horn; the monoceros. This name is often applied to the rhinoceros. You may not know that some rhinoceros have one horn and that is what they were referring to most likely.

Critics of the Bible would state that there are inconsistencies in the Bible. When examining the Bible there are 66 books of the Bible. With over 700,000 words in the Bible with the King James version. Out of that, there are several passages that don't match up.

Matthew 21:19: And when he saw a fig tree in the way, he came to it, and found nothing thereon, but leaves only, and said unto it, Let no fruit grow on thee henceforward forever. And presently the fig tree withered away. It happened immediately, and the disciples were amazed. Vs. *Mark 11:20*: And in the morning, as they passed by, they saw the fig tree dried up from the roots. The dried-up fig tree was discovered the following morning.

Mark 6:53: When they had crossed over, they landed at Gennesaret and anchored there.

Jesus and the disciples went to Gennesaret. Vs.John 6:24-25: Once the crowd realized that neither Jesus nor his disciples were there, they got into the boats and went to Capernaum in search of Jesus. When they found him on the other side of the lake, they asked him, "Rabbi, when did you get here?" Jesus and the disciples went to Capernaum.

Matthew 27:28: They stripped him and put a scarlet robe on him.

Jesus is given a scarlet robe. Vs *John 19:2:* The soldiers twisted together a crown of thorns and put it on his head. They clothed him in a purple robe. Jesus is given a purple robe.

Matthew 28:1: After the Sabbath, at dawn on the first day of the week, Mary Magdalene and the other Mary went to look at the tomb. Mary Magdalene and the other Mary vs.*Mark 16:1*: When the Sabbath was over, Mary Magdalene, Mary the mother of James, and Salome bought spices so that they might go to anoint Jesus' body.

The two Marys, plus a third person, Salome, vs Luke 24:10: When they came back from the tomb, they told all these things to the Eleven and to all the others. It was Mary Magdalene, Joanna, Mary the mother of James, and the others with them who told this to the apostles. The two Marys, Joanna, and "the others." vs *John 20:1:* Early on the first day of the week, while it was still dark, Mary Magdalene went to the tomb and saw that the stone had been removed from the entrance. Only Mary Magdalene

There are inconsistencies, but this does not mean it's not true. If we were in a classroom and we were each taking notes, we would have different notes. If someone was making up the Bible then there wouldn't be any inconsistencies in the Bible. Because there this gives the Bible more authenticity. There are four Gospels: Matthew, Mark, Luke, and John. Skeptics would say that Matthew, John, and Luke copied from Mark. If that were the case, there would be all kinds of inconsistencies; each of these Gospels adds details that the others didn't.

There are numerical inconsistencies 2 Samuel 10:18 the number of Chariots number killed by David as 700, while I Chronicles 19:18 lists 7,000 men killed in the incident. 2 Chronicles 36:9 where Jehoiachin is listed as eight years old. In 2 Kings 24:8 it lists him at eighteen.

The New York Times is one of the best-known newspapers. On average, it has 150,000 words in a given newspaper. The Bible has about 5 times as many words. On Sunday, September 3 there were 4 mistakes. In our modern times, we have spell check, and we have grammatical check. We have people who have more schooling and are able to pick up mistakes. In modern times, we still make mistakes. Then why can't 2000 years ago they could have made mistakes?

Sometimes we may also think there are mistakes when there aren't any mistakes. The Bible was made up of different languages, but a lot of the Bible was Hebrew. In the original Hebrew text, there were two letters that were in the text that did not make sense at the time Aleph and Tav. Now this was the original Hebrew. With the original Hebrew, there were more pictures than the formation of letters. The letters had meaning. I have listed these two letters down below, along with their meaning.

Aleph		Ox / strength / leader
Tav	†	Mark / sign / covenant

The extra letters were found in *John 1:1.* **In the beginning was the Word, and the Word was with God, and the word was God.** The significance of these letters is that they are the first letter of the Hebrew alphabet and the last letter. Through the Bible, there were several references to the first and last. You may be familiar with the term Alpha and Omega. *Revelation 22:13* **"I am the Alpha and the Omega, the first and the last, the beginning and the end."** Another reference to the Alpha and the Omega. *Revelation 21:6-7* **Then He said to me, "It is done I am the Alpha and the Omega, the beginning and the end I will give to the one who thirsts from the spring of the water of life without cost. "He who overcomes will inherit these things, and I will be his God and he will be My son.**

Let's examine the two letters. The first letter is Aleph, which is the first letter of the alphabet. If you see it's like a bull head. Now, modern-day Hebrew is different from back then. The Aleph letter a lot of times, was a reference to God. The Aleph character you could see is a head. The character means leader. You could also make a case for Aelph being the Old Testament, with the bull being sacrificed.

The second letter, the Tav, is the last letter of the alphabet. The letter is associated with a mark or covenant. You have to remember that the Hebrew letters were made before Jesus was born. You could see the sign is a Christian cross. Although there are a few passages in the Bible that have the Tav Ezekiel and Job. You could see this letter as being associated with Jesus. As a Christian, there is no bigger covenant bigger than Jesus. *Luke*

22:20 NLV 20 **After supper, he took another cup of wine and said, "This cup is the new covenant between God and his people—an agreement confirmed with my blood, which is poured out as a sacrifice for you.**

In *John 3:16* it further spells out what the covenant is 20**For God so loved the world that he gave his one and only Son, that whoever believes in him shall not perish but have eternal life.**

Conclusion

We took a lot of different things in the Bible. In the Bible, it talked about things we didn't know until they were scientifically proven years later. Like running water when you have unclean hands. This won't be found until 2000 years later. Or circumcision, where the Bible talks about the one scientifically best date to circumcise. It is the date when the babies start to have a blood clot. It also talked about Earth being a sphere and describes possible dinosaurs. We also looked separately at people when they have leprosy. If this could have been done during the Middle Ages, it would have saved a lot of people.

We also looked at something that originally focused on mistakes. There were two letters in the original Hebrew that didn't make sense. The alph and the Tev. These were the beginning letter of the alphabet and the end. One letter representing God the other letter representing Jesus Christ.

We also looked for inconsistencies in the Bible. Some skeptics would think this disproves the Bible. When you talk with police detectives. You are going to have different people have a different perspective. We looked at the New York Times on a given day. The Bible had fewer mistakes than the New York Times. In modern times, we have spell check, but we still make grammatical mistakes. The paper that it was put on was very expensive to put on.

The Bible also talked about current times, including airplanes, selfies, disobedience, and also talked about nuclear bombs. We also talked about the inconsistencies with the Bible. It is really overemphasized. This should give you more faith in the Bible. In the next section, we will talk about the proof of Jesus Christ.

3. A WHAT PROOF IS THERE JESUS CHRIST FROM A BIBLICAL PERSPECTIVE?

There were many things that prove Jesus Christ. We will talk about the prophecies that ended up coming in true in Jesus. In another part, we will further examine proof of Jesus Christ through non-Christian historians at the time. Prophecies are predictions that were hundreds of years before Jesus Christ was born. They have dated the Old Testament, which was carbon-dated before Jesus Christ. We will discuss some of the prophecies that we talked about in the Old Testament (before Jesus Christ) and what was fulfilled in Jesus time.

When you translate the first names of the Bible it translates into a message. They say that the Genesis book was completed 1600 years before Jesus came. All these names were in Genesis.

1. God **God**
2. Adam **Man**
3. Seth**is appointed**
4. Enosh**a mortal man of**
5. Kenan **sorrow is born!**
6. Jared **The GLORY of God**
7. Mahalalel**Shall come Down**
8. Enoch**Instructing that**
9. Methuselah **His death shall bring**
10. Lamech**Those in despair**
11. Noah **Comfort and Rest**

Again, this goes against points to Jesus Christ will come down from Earth, instructing us on how to live and that God plan was he was always going to die.

1. Abraham's seed, David seed fulfilled the Old Testament. In *Genesis 1:18, s***ince Abraham is destined to become a great and mighty nation, and all the nations of the earth will be blessed through him?** That was in the Old Testament. It was fulfilled in the New Testament

A. Abraham Seed David seed fulfilled the New Testament In *Matthew 1:1 (Amp)* 1 **The record of the genealogy of Jesus the Messiah, the son (descendant) of David, the son (descendant) of Abraham.**

2. (Son of God Old Testament)*Psalm 2 NLV* **What fools the nations are to rage[a] against the Lord! How strange that men should try to outwit God! 2 For a summit conference of the nations has been called to plot against the Lord and his Messiah, Christ the King. 3"Come, let us break his chains," they say, "and free ourselves from all this slavery to God."4 But God in heaven merely laughs! He is amused by all their puny plans. 5 And then in fierce fury he rebukes them and fills them with fear.6 For the Lord declares, "This is the King of my choice, and I have enthroned him in Jerusalem, my holy city."7His chosen one replies, "I will reveal the everlasting purposes of God, for the Lord has said to me, 'You are my Son. This is your Coronation Day.* Today I am giving you your glory.'" 8 "Only ask and I will give you all the nations of the world. 9 Rule them with an iron rod; smash them like clay pots!"**

A.(Son of God New Testament) In *Matthew 3 16:17 (NLT)* 16 **After his baptism, as Jesus came up out of the water, the heavens were opened[g], and he saw the Spirit of God descending like a dove and settling on him. 17 And a voice from heaven said, "This is my dearly loved Son, who brings me great joy."**

3.(His resurrection Old Testament) Psalm 2 1-2 *(NLV)* **For you will not leave my soul among the dead[or allow your holy one to rot in the grave. You will show me the way of life, granting me the joy of your presence and the pleasures of living with you forever.**

A.(His resurrection New Testament) *Acts 26:22 (NLV)* **But God has protected me right up to this present time so I can testify to everyone, from the least to the greatest. I teach nothing except what the prophets and Moses said would happen— 23 that the Messiah would suffer and be the first to rise from the dead, and in this way announce God's light to Jews and Gentiles alike."**

4.(Hands and Feet pierced Old Testament) *Psalm 22:1* **My God, my God, why have you abandoned me? Why are you so far away when I groan for help? Everyday I call to you, my God, but you do not answer. Every night I lift my voice, but I find no relief. Yet you are holy, enthroned on the praises of Israel. 4 Our ancestors trusted in you, and you rescued them. 5 They cried out to you and were saved. They trusted in you and were never disgraced. 6 But I am a worm and not a man. I am scorned and despised by all! 7 Everyone who sees me mocks me. They sneer and shake their heads, saying, 8 "Is this the one who relies on the Lord? Then let the Lord save him! If the Lord loves him so much, let the Lord rescue him!" 9 Yet you brought me safely from my mother's womb and led me to trust you at my mother's breast. 10 I was thrust into your arms at my birth. You have been my God from the moment I was born. 11 Do not stay so far from me, for trouble is near, and no one else can help me. 12 My enemies surround me like a herd of bulls; fierce bulls of Bashan have hemmed me in! 13 Like lions, they open their jaws against me, roaring and tearing into their prey. 14 My life is poured out like water, and all my bones are out of joint. My heart is like wax, melting within me. 15 My strength has dried up like sunbaked clay. My tongue sticks to the roof of my mouth. You have laid me in the dust and left me for dead. 16 My enemies surround me like a pack of dogs; an evil gang closes in**

on me. They have pierced my hands and feet. 17 I can count all my bones. My enemies stare at me and gloat.18 They divide my garments among themselves and throw dice[for my clothing O Lord, do not stay far away! 19 You are my strength; come quickly to my aid. 20 Save me from the sword; spare my precious life from these dogs. 21 Snatch me from the lion's jaws and from the horns of these wild oxen. 22I will proclaim your name to my brothers and sisters. I will praise you among your assembled people.

A.Hands of feet pierced the New Testament *John 20:26-27 AMP* 26 **Eight days later His disciples were again inside the house, and Thomas was with them. Jesus came, through the doors had been barred, and stood among them and said, "Peace to you." 27 Then He said to Thomas, "Reach here with your finger, and see My hands; and put out your hand and place it in My side. Do not be unbelieving, but [stop doubting and] believe." 28 Thomas answered Him, "My Lord and my God!" 29 Jesus said to him, "Because you have seen Me, do you now believe? Blessed [happy, spiritually secure, and favored by God] are they who did not see [Me] and yet believed [in Me].**"*John 19 34-36 AMP* 34 **But one of the soldiers pierced His side with a spear, and immediately blood and water came [flowing] out. 35 And he (John, the eyewitness) who has seen it has testified, and his testimony is true; and he knows that he is telling the truth, so that you also [who read this] may believe. 36 For these things took place to fulfill the Scripture, "Not a bone of His shall be broken." 37 And again another Scripture says, "They shall look at Him whom they have pierced."**

5. Born of a Virgin Old Testament (Isaiah 7:14 NLT) **All right then, the Lord himself will give you the sign. Look! The virgin will conceive a child! She will give birth to a son and will call him Immanuel (which means 'God is with us'). 15 By the time this child is old enough to choose what is right and reject what is wrong, he will**

be eating yogurt and honey. 16 For before the child is that old, the lands of the two kings you fear so much will both be deserted.

B. Born of a Virgin New Testament *Matthew 20-22* **As he considered this, an angel of the Lord appeared to him in a dream. "Joseph, son of David," the angel said, "do not be afraid to take Mary as your wife. For the child within her was conceived by the Holy Spirit. 21 And she will have a son, and you are to name him Jesus, for he will save his people from their sins."23 All of this occurred to fulfill the Lord's message through his prophet:24 "Look! The virgin will conceive a child! 25 She will give birth to a son, and they will call him Immanuel, which means 'God is with us.'" 26.Full of Wisdom and power Old Testament.** *Isaiah 11:1-4 NLV)* **Out of the stump of David's family will grow a shoot—yes, a new Branch bearing fruit from the old root, and the Spirit of the Lord will rest on him—the Spirit of wisdom and understanding, the Spirit of counsel and might, the Spirit of knowledge and the fear of the Lord. He will delight in obeying the Lord. He will not judge by appearance nor make a decision based on hearsay. He will give justice to the poor and make fair decisions for the exploited.**

6. Full of wisdom and power, New Testament (Hebrews 1:4 NLV) **Long ago God spoke many times and in many ways to our ancestors through the prophets. 2 And now in these final days, he has spoken to us through his Son. God promised everything to the Son as an inheritance, and through the Son, he created the universe. 3 The Son radiates God's own glory and expresses the very character of God, and he sustains everything by the mighty power of his command. When he had cleansed us from our sins, he sat down in the place of honor at the right hand of the majestic God in heaven. 4 This shows that the Son is far greater than the angels, just as the name God gave him is greater than their names**

7. The Meek Servant, Old Testament (Isaiah 42 1:7) **"Behold, My Servant, whom I uphold; My Chosen One in whom My soul delights.**

I have put My Spirit upon Him. He will bring forth justice to the nations."He will not call out or shout aloud, nor make His voice heard in the street. 3 "A broken reed He will not break And a dimly burning wick He will not extinguish [He will not harm those who are weak and suffering]; He will faithfully bring forth justice.4 "He will not be disheartened or crushed [in spirit];[He will persevere] until He has established justice on the earth; And the coastlands will wait expectantly for His law."5 This is what God the Lord says, He who created the heavens and stretched them out, Who spread out the earth and its produce, Who gives breath to the people on it And spirit to those who walk on it 6 "I am the Lord, I have called You (the Messiah) in righteousness [for a righteous purpose], I will also take You by the hand and keep watch over You, And I will appoint You as a covenant to the people [Israel], As a light to the nations (Gentiles),7 To open the eyes of the blind, To bring out prisoners from the dungeon. And those who sit in darkness from the prison.

A.The Meek Servant New Testament *Matthew 12 18-19* **"Behold, My Servant whom I have chosen; My Beloved in whom My soul is well-pleased; I will put My Spirit upon Him, And He will proclaim justice to the nations.19 "He will not quarrel, nor cry out loudly; Nor will anyone hear His voice in the streets.**

8. Rejected by his people Crucified old testament *Isaiah 53 11-12* Amp **As a result of the anguish of His soul, He shall see it and be satisfied; By His knowledge [of what He has accomplished] the Righteous One, My Servant, shall justify the many [making them righteous—upright before God, in right standing with Him], For He shall bear [the responsibility for] their sins.12 Therefore, I will divide and give Him a portion with the great [kings and rulers]. He shall divide the spoils with the mighty, Because He [willingly] poured out His life to death, And was counted among the transgressors. Yet, He Himself bore and took away the sin of many, And interceded [with the Father] for the transgressors.**

A.Rejected by his people New Testament *Mark 14 61-65* Amp 61 **But**
Jesus kept silent and gave no answer at all. Again the high priest was
questioning Him, and saying to Him, "Are You the Christ (the
Messiah, the Anointed), the Son of the Blessed One?" 62 **Jesus said, "I**
am; and you will [all] see the Son of Man seated [with authority] at
the right hand of Power (the Father), and coming with the clouds of
heaven." 63 **Then, tearing his robe [to express his indignation], the**
high priest said, "What further need do we have of witnesses? 64 **You**
have heard the blasphemy [that is, His claim to be the Son of God].
What is your decision?" And they all condemned Him to be [guilty
and] deserving of death. 65 **And some began to spit on Him, and to**
blindfold Him, and to beat Him with their fists, and to say to Him,
"Prophesy [by telling us who hit you]!" Then the officers took
custody of Him and struck him in the face.

9. New Covenant old testament *Jeremiah 31: 31-34 31* **"Behold, the**
days are coming," says the Lord, "when I will make a new covenant
with the house of Israel (the Northern Kingdom) and with the house
of Judah (the Southern Kingdom), 32 **not like the covenant which I**
made with their fathers in the day when I took them by the hand to
bring them out of the land of Egypt, My covenant which they broke,
al-though I was a husband to them," says the Lord. 33 **"But this is the**
covenant which I will make with the house of Israel after those
days," says the Lord, "I will put My law within them, and I will write
it on their hearts; and I will be their God, and they will be My people.
34 **And each man will no longer teach his neighbor and his brother,**
saying, 'Know the Lord,' for they will all know Me [through
personal experience], from the least of them to the greatest," says the
Lord. "For I will forgive their wickedness, and I will no longer
remember their sin."

A.New Covenant new testament *Matthew 26:26 31* **"Behold, the days**
are coming," says the Lord, "when I will make a new covenant with
the house of Israel (the Northern Kingdom) and with the house of

**Judah (the Southern Kingdom), 32 not like the covenant which I
made with their fathers in the day when I took them by the hand to
bring them out of the land of Egypt, My covenant which they broke,
though I was a husband to them," says the Lord. 33 "But this is the
covenant which I will make with the house of Israel after those
days," says the Lord, "I will put My law within them, and I will write
it on their hearts; and I will be their God, and they will be My people.
34 And each man will no longer teach his neighbor and his brother,
saying, 'Know the Lord,' for they will all know Me [through
personal experience], from the least of them to the greatest," says the
Lord. "For I will forgive their wickedness, and I will no longer
remember their sin."**

10. Born in Bethlehem New testament *Micah 5: 2-4 9 Amp* **"But as for
you, Bethlehem Ephrathah, Too little to be among the clans of
Judah; From you One shall come forth for Me [who is] to be Ruler
in Israel, His goings forth (appearances) are from long ago, From
ancient days." 3 Therefore, He will give them up until the time.
When she who is in labor has given birth to a child. Then what is
left of His kinsmen? Shall return to the children of Israel. 4 And He
shall stand and shepherd and guide His flock. In the strength of the
Lord, In the majesty of the name of the Lord His God; And they shall
dwell [secure in undisturbed peace], because at that time He shall be
great [extending His authority][Even] to the ends of the earth.**

A. Born in Bethlehem, New Testament, *Luke 2:1 Amp* **Now when
Jesus was born in Bethlehem of Judea in the days of Herod the king
(Herod the Great), magi (wise men) from the east came to Jerusalem,
asking, 2 "Where is He who has been born King of the Jews? For we
have seen His star in the east and have come to worship Him.**

11.Sold for pieces of silver Old Testament Zechariah Amp 11:12-13 ***12 I
said to them, "If it seems good to you, give me my wages; but if not,
do not." So they weighed out [a]thirty pieces of silver as my wages.
13 Then the Lord said to me, "Throw it to the potter [as if to the***

dogs]—that magnificent sum at which I am valued by them!" So I took the thirty pieces of silver and threw them to the potter in the house of the Lord.

A.Sold for pieces of Silver *Matthew 26 14-15 Amp* 14 **Then one of the twelve [disciples], who was called Judas Iscariot, went to the chief priests** 15 **and said, "What are you willing to give me if I hand Jesus over to you?" And they weighed out thirty pieces of silver. 16 And from that moment, Judas began looking for an opportune time to betray Jesus.**

12. Piercing of His Body, Old Testament, *Zechariah 12:10 Amp 10* **"I will pour out on the house of David and on the people of Jerusalem, the Spirit of grace (unmerited favor) and supplication. And they will look at Me, whom they have pierced; and they will mourn for Him as one mourns for an only son, and they will weep bitterly over Him as one who weeps bitterly over a firstborn.**

A.Piercing of his body New Testament *John 19:34 Amp 34* **But one of the soldiers pierced His side with a spear, and immediately blood and water came [flowing] out.** Our Sins purged the Old Testament *Malachi 3:1-3 Amp* **"Behold, I am going to send My messenger, and he will prepare and clear the way before Me. And the Lord [the Messiah], whom you seek, will suddenly come to His temple; the Messenger of the covenant, in whom you delight, behold, He is coming," says the Lord of hosts.** 2 **But who can endure the day of His coming? And who can stand when He appears? For He is like a refiner's fire and like a launderer's soap [which removes impurities and uncleanness].** 3 **He will sit as a refiner and purifier of silver, and He will purify the sons of Levi [the priests], and refine them like gold and silver, so that they may present to the Lord [grain] offerings in righteousness.**

A. Our sins are purged. New Testament *1 Peter: 18 Amp* **"Behold, I am going to send My messenger, and he will prepare and clear the way before Me. And the Lord [the Messiah], whom you seek, will suddenly come to His temple; the Messenger of the covenant, in**

whom you delight, behold, He is coming," says the Lord of hosts. 2
But who can endure the day of His coming? And who can stand
when He appears? For He is like a refiner's fire and like a launderer's
soap [which removes impurities and uncleanness]. 3 He will sit as a
refiner and purifier of silver, and He will purify the sons of Levi [the
priests], and refine them like gold and silver, so that they may
present to the Lord [grain] offerings in righteousness

The biggest prophecy of all them all was the prediction of the birth of Christ. This was predicted by Daniel, one of the major prophets who foretold the downfall of several empires of the day.

14. Time of his birth Old testament *Daniel 9 24-27 AMP 24* **"Seventy**
weeks [of years, or 490 years] have been decreed for your people and
for your holy city (Jerusalem), to finish the transgression, to make
an end of sins, to make atonement (reconciliation) for wickedness,
to bring in everlasting righteousness (right-standing with God), to
seal up vision and prophecy and prophet, and to anoint the Most
Holy Place. 25 So you are to know and understand that from the
issuance of the command to restore and rebuild Jerusalem until [the
coming of] the Messiah (the Anointed One), the Prince, there will be
seven weeks [of years] and sixty-two weeks [of years]; it will be built
again, with [a city] plaza and moat, even in times of trouble. 26 Then
after the sixty-two weeks [of years] the Anointed One will be cut off
[and denied His Messianic kingdom] and have nothing [and no one
to defend Him], and the people of the [other] prince who is to come
will destroy the city and the sanctuary. Its end will come with a
flood; even to the end, there will be war; desolations are determined.
27 And he will enter into a binding and irrevocable covenant with the
many for one week (seven years), but in the middle of the week he
will stop the sacrifice and grain offering [for the remaining three
and one-half years]; and on the wing of abominations will come one
who makes desolate, even until the complete destruction, one that is
decreed, is poured out on the one who causes the horror."

Let's break this down a little bit more.

457 27 AD31 AD34 AD

Decree toJesus wasJesus was Stephen was

RebuildBaptizedCrucifiedstoned

JerusalemDan 9:25Dan 9:26Acts 7:59

A.Time of his birth New Testament *Luke 2 1-2 Amp* **Now in those days a decree went out from [the emperor] Caesar Augustus, that all the inhabited world (the Roman Empire) should be registered [in a census]. 2 This was the first census taken while Quirinius was governor of Syria**

This is what Jesus said about when back or when it is the end of days In Matthew 24 1- 25 **Jesus left the temple area and was going on His way when His disciples came up to Him to call His attention to the [magnificent and massive] buildings of the temple. 2 And He said to them, "Do you see all these things? I assure you and most solemnly say to you, not one stone here will be left on another, which will not be torn down."3 While Jesus was seated on the Mount of Olives, the disciples came to Him privately, and said, "Tell us, when will this [destruction of the temple] take place, and what will be the sign of Your coming, and of the end (completion, consummation) of the age?"While Jesus was seated on the Mount of Olives, the disciples came to Him privately, and said, "Tell us, when will this [destruction of the temple] take place, and what will be the sign of Your coming, and of the end (completion, consummation) of the age?"4 Jesus answered, "Be careful that no one misleads you [deceiving you and leading you into error]. 5 For many will *come* in My name [misusing it, and appropriating the strength of the name which belongs to Me], saying, 'I am the Christ (the Messiah, the Anointed),' and they will mislead many. 6 You will *continually* hear of wars and rumors of wars. See that you are not frightened, for *those things* must take place, but that is not yet the end [of the age].**

3. B WHAT IS PROOF OF JESUS CHRIST FROM A NON-BIBLICAL PERSPECTIVE?

You have to understand that Jesus Christ lived over 2000 years ago. That many years, but yet so many years ago, but yet so many copies exist. There are 5,366 manuscripts of Jesus Christ. (These are from the 1st and 2nd century). There are 643 copies of Homer Iliad. And there are only 10 copies of Julius Caesar's Gallic Wars. There are so many records of Jesus Christ.

We talked about the Bible proving Jesus Christ. This section will take a look at historians who talk about Jesus. It will also look at historians of the day that support the Bible or things that were brought up in the Bible.

The first historical person we want to talk about is Flavius Josephus. He was Romano- Jewish leader in the first century. In his book Antiques of the Jews, which was written in 93-94. This is what he wrote in the Antiques of the Jews Book 18 Chapter 3: This is what was written:

[About this time there lived Jesus, a wise man, if indeed one ought to call him a man. For he was one who performed surprising deeds and was a teacher of such people as accept the truth gladly. He won over many Jews and many of the Greeks. He was the Christ. And when, upon the accusation of the principal men among us, Pilate had condemned him to a cross, those who had first come to love him did not cease. He appeared to them spending a third day restored to life, for the prophets of God had foretold these things and a thousand other marvels about him. And the tribe of the Christians, so called after him, has still to this day not disappeared.]

Josephus also wrote about John the Baptist, Antiquities of the Jews, Book 18, Chapter 5

Now some of the Jews thought that the destruction of Herod's army came from God, and that very justly, as a punishment of what he did against John, that was called the Baptist: for Herod slew him, who was a good man... Herod, who feared lest the great influence John had over the people might put it into his power and inclination to raise a rebellion... Accordingly, he was sent a prisoner, out of Herod's suspicious temper, to Machaerus, the castle I mentioned before, and was put there to death.

Josephus also talked about Jesus' brother James. In the Bible it talked about Jesus having a brother. It is not clear whether it was a stepbrother or a cousin. In the Bible when they did talk about James, Jesus brother, he is referred to as James the lesser, James, son of Alphaeus, bishop of Jerusalem, and also brother of Jesus

In Matthew 13 55 **"Isn't this the carpenter's son? Isn't his mother's name Mary, and aren't his brothers James, Joseph, Simon and Judas? Aren't all his sisters with us? Where then did this man get all these things?" And they took offense at him.**

In Mark 15:40 40 **Now some women also were watching from a distance, among whom were Mary Magdalene, and Mary the mother of James the Less and of Joses, and Salome.** 41 **When Jesus was in Galilee, they used to accompany him and minister to Him; and there were also many other women who came up with Him to Jerusalem.**

In the Antiques of the Jews Book 20 Chapter 9, he talked about Jesus and his brother James. This is what he wrote *Festus was now dead, and Albinus was but upon the road; so he assembled the sanhedrin of judges, and brought before them the brother of Jesus, who was called Christ, whose name was James, and some others; and when he had formed an accusation against them as breakers of the law, he delivered them to be stoned: but as for those who seemed the most equitable of the citizens, and such as were the most uneasy at the breach of the laws, they disliked what was done; they also sent to the king, desiring him to send to Ananus that he should act so no more, for that what he had already done was not to be justified; nay, some of them went also to meet Albinus, as he was upon his journey from Alexandria, and informed him that it was not lawful for Ananus to assemble a sanhedrin without his consent. Whereupon Albinus complied with what they said, and wrote in anger to Ananus, and threatened that he would bring him to punishment for what he had done; on which king Agrippa took the high priesthood from him, when he had ruled but three months, and made Jesus, the son of Damneus, high priest.*

Tacitus was a Roman senator and historian from 55-120. He wrote the Annals passage book 15 chapter 44 *Consequently, to get rid of the report, Nero fastened the guilt and inflicted the most exquisite tortures on a class hated for their abominations, called Christians by the populace. Christus, from whom the name had its origin, suffered the extreme penalty during the reign of Tiberius at the hands of one of our procurators, Pontius Pilatus, and a most mischievous superstition, thus checked for the moment, again broke out not only in Judæa, the first source of the evil, but even in Rome, where all things hideous and shameful from every part of the world find their centre and become popular. Accordingly, an arrest was first made of all who pleaded guilty; then, upon their information, an immense multitude was convicted, not so much of the crime of firing the city, as of hatred against mankind.*

In the passage, he talks about Christ, who suffered an extreme penalty, and Pontius Pilate who, in the Bible killed Jesus. You could see that Tacitus was not a "fan of Christians," he called them an abomination. If there was no Jesus, he certainly would not lie about that. He had no motive to lie.

Suetonius was an historian between 69 AD and 122 AD. He talks about Christians and Jesus Christ. "Since the Jews constantly made disturbances at the instigation of Chrestus (Jesus Christ), he expelled them from Rome." This matches up to what *Acts 18:2 AMP* 2 **There he met a Jew named Aquila, a native of Pontus, who had recently come from Italy with his wife, Priscilla, because [the Roman Emperor] Claudius had issued an edict that all the Jews were to leave Rome. Paul went to see them.**

Suetonius also talked about Nero. If you remember history, Nero went after Christians and fed them to lions. It wasn't talked about in the Bible. "*During his reign many abuses were severely punished and put down, and no fewer new laws were made: a limit was set to expenditures; the public banquets were confined to a distribution of food; the sale of any kind of cooked viands in the taverns was forbidden, with the exception of pulse and vegetables, whereas before every sort of dainty was exposed for sale. Punishment was inflicted on the Christians, a class of men given to a new and mischievous superstition*. Suetonius confirms that followers of Christ. He also confirmed that this was a new movement. So it would have been something different that the Jewish religion was doing. Keep in mind that Jewish people were practicing their religion for almost 1,000 years before Christ came into this world.

Lucian was a playwright who talked about Christians. In his play The Passing of Peregrinus or The Death of Peregrinus. This is what he wrote in this play about Christians: The Christians, you know, worship a man to this day—the distinguished personage who introduced their novel rites, and was crucified on that account. ... *You see, these misguided creatures start with the general conviction that they are immortal for all time, which explains their contempt of death and voluntary self-devotion which are so common among them; and then it was impressed on them by their original lawgiver that they are all brothers, from the moment that they are converted, and deny the gods of Greece, and worship the crucified sage, and live after his la*ws. All this they take quite faithfully, with the result that they despise all worldly goods alike, regarding them merely as common property.

From the text, you could see how it describes Christ as the leader of Christians. He states how important it is to the religion. How Christians thought of the afterlife, which was a departure from Roman life. He also stated that Christians did not mind dying as they knew they were going to heaven.

There was an important inscription found in Nazareth in 1878. It was a decree that Claudius (41-78 AD). The decree follows as

EDICT OF CAESAR

It is my decision [concerning] graves and tombs—whoever has made them for the religious observances of parents, or children, or household members—that these remain undisturbed forever. But if anyone legally charges that another person has destroyed, or has in any manner extracted those who have been buried, or has moved with wicked intent those who have been buried to other places, committing a crime against them, or has moved sepulcher-sealing stones, against such a person, I order that a judicial tribunal be created, just as [is done] concerning the gods in human religious observances, even more so will it be obligatory to treat with honor those who have been entombed. You are absolutely not to allow anyone to move [those who have been entombed]. But if [someone does], I wish that [violator] to suffer capital punishment under the title of tomb-breaker.

One of the arguments that people had was that Jesus' tomb was moved. This challenges that because not only would they have to move the stone. Whoever would have stolen the body would have risked death if they were to move Jesus' body.

This is what Jesus said about when back or when it is the end of days In Matthew 24 1- 25 **Jesus left the temple area and was going on His way when His disciples came up to Him to call His attention to the [magnificent and massive] buildings of the temple. 2 And He said to them, "Do you see all these things? I assure you and most solemnly say to you, not one stone here will be left on another, which will not be torn down."3 While Jesus was seated on the Mount of Olives, the disciples came to Him privately, and said, "Tell us, when will this [destruction of the temple] take place, and what will be the sign of Your coming, and of the end (completion, consummation) of the age?"While Jesus was seated on the Mount of Olives, the disciples came to Him privately, and said, "Tell us, when will this [destruction of the temple] take place, and what will be the sign of Your coming, and of the end (completion, consummation) of the age?"4 Jesus answered, "Be careful that no one misleads you [deceiving you and leading you into error]. 5 For many will *come* in My name [misusing it, and appropriating the strength of the name which belongs to Me], saying, 'I am the Christ (the Messiah, the Anointed),' and they will mislead many. 6 You will *continually* hear of wars and rumors**

of wars. See that you are not frightened, for *those things* must take place, but that is not yet the end [of the age].

Conclusion

We looked at how many records. It is believed that Homer was a real person, yet there are about

600 copies of that. The Bible has over 5,000 historic copies.

We took several historical records. We took a look at Flavius Josephus, who talked about Jesus being a leader of the Christians. Josephus was a Roman he would be objective.

Josephus also gave us details which are listed in the Bible such as John the Baptist getting killed by Herod, and also it talked about Jesus Christ having a younger brother, James.

We also had historians Tacitus, Suetonius, and Lucian the playwright. These people all matched up something to the Bible. Lucian talked about the expulsion of Jews from Rome after Christ's death. Tactus, which confirmed that Christ was executed by Pontius Pilate. Suetonius confirmed that there were followers of Christ, and they were persecuted.

4. WHY DID GOD PUT JESUS EARLIER IN THIS WORLD?

If God knew that he would need Jesus, why not put Jesus in this world right away? If we need Jesus, then why not put them on Earth? Jesus bridges the gap between man and God. He brought the light and preached about God. There is no more important person who has lived on Earth than Jesus.

God is so smart, you've to remember that. With God there is always order. There was order with creation. We had to have light from darkness, then the earth, then the animals. We started with a new beginning with God. We had a huge flood from there, we had Noah. Noah was a good man. From there, we had Abraham, the father of many nations. This was the seed for the father of Israel. If Jesus Christ had come into the world at this time, there would be few to preach to.

Next in the Bible Moses came into the Bible. Moses was told by God to lead the people to the promised land. Before he led them to the promised land. God showed them ten plagues before people believed that Moses had God on their side. They say it was ten plagues because the Israeli people worshipped 10 GodS. God wanted these people to be set apart from the rest of the world. Along the way, Moses went up on the mountain. God made the Ten Commandments on the stone tablets. That started a relationship between God and his people, instead of people, doing what they wanted to do. They started to do what God wanted. Along the way God provided for the Israeli people by providing food, they provided protection for people, they provided protection for his people.

They started to include more laws on how to act or treat. They say there were more than 300 laws. Why not insert Jesus then? We could argue that the people's hearts were not right to accept Jesus. These people saw that the dead sea split so they could walk into it. They see the 10 plagues of Egypt.

Yet they still worshipped a golden calf while Moses was on Mount Sinai. So they weren't ready to receive Christ.

So the next period was the king's. The prophet Samuel anointed Saul. (Samuel was a great prophet and had a lot of influence in Israel. *1 Samuel 9 15-17* **15 Now the Lord had told Samuel the previous day, 16 "About this time tomorrow I will send you a man from the land of Benjamin. Anoint him to be the leader of my people, Israel. He will rescue them from the Philistines, for I have looked down on my people in mercy and have heard their cry."17 When Samuel saw Saul, the Lord said, "That's the man I told you about! He will rule my people."The Israeli people wanted a king because other lands had kings. Saul was part of the Benjamin tribe. They were the first tribe to get land. Joshua 18: 11 The first allotment of land went to the clans of the tribe of Benjamin. The tribe of Benjamin was one of the smallest tribe of the tribes of Israel.**

Samuel started to lose favor with God. *1 Samuel 13-14* **13 "How foolish!" Samuel exclaimed. "You have not kept the command the Lord your God gave you. Had you kept it, the Lord would have established your kingdom over Israel forever. 14 But now your kingdom must end, for the Lord has sought out a man after his own heart. The Lord has already appointed him to be the leader of his people, because you have not kept the Lord's command."**

Saul tries to kill David several times. David had a favor from God. So Saul was Jealous. In *1 Samuel 19 9 -10* **9 But one day when Saul was sitting at home, with spear in hand, the tormenting spirit from the Lord suddenly came upon him again. As David played his harp, 10 Saul hurled his spear at David. But David dodged out of the way, and leaving the spear stuck in the wall, he fled and escaped into the night.**

David wins favor when he defeats Goliath. *1 Samuel 17 48 -51* **48 As Goliath moved closer to attack, David quickly ran out to meet him.49 Reaching into his shepherd's bag and taking out a stone, he**

hurled it with his sling and hit the Philistine in the forehead. The stone sank in, and Goliath stumbled and fell face down on the ground.50 So David triumphed over the Philistine with only a sling and a stone, for he had no sword. 51 Then David ran over and pulled Goliath's sword from its sheath. David used it to kill him and cut off his head.

David then becomes king. He was seen as a great conqueror, having won many battles. He was also anointed by God to be the leader. One of the most important battles that he won was that he was able to get the ark of the convenient back. David sees Bathsheba taking a bath, David impregnates Bathsheba. To cover his sin up. He tried to get Uriah (Bathsheba) to come back and tried to get them pregnant when that didn't happen. He got Uriah to the front line during the war. David hopes that he will die. When Uriah dies, he personally tells Bathsheba, and they end up getting married. Nathan the prophet says that God was very displeased with this.

Well, why didn't Jesus come then? I believe it was for a couple of reasons. One of the reasons that God did not want to send Jesus during the king's period is that he wanted to show how imperfect human kings are. They start off as good kings, but sin comes in and makes them not good kings. In the case of Saul, they did not obey God. When Saul was no longer favored anymore, he tried to kill David. Davide impregnated another person's wife. Then try to kill the men. Solomon was David's successor. He was said to be the wisest man who ever lived. He is the author of much of the Proverbs. Solomon is said to have had over 500 women. One of those women got Solomon to worship other GodS.

During this period of the Kings. There would be a King who would be elevated. Why not bring Jesus in this time? God wanted us not to put faith in men when it comes to our king. That every great man had sin, and God had to show us that we need a king without error, we need Jesus. A king may be as smart as Solomon, but they can still sin. They may be spiritually anointed like Saul, but sin can still come in. They could be brave and victorious like David. As victorious as David, he had sinned.

Conclusion: We explored why Jesus came when he came. We had to know what good leadership was. We had to choose people in order for God's seed to be brought into the world. We had to have laws to understand how much we have fallen from what God wants us to be. Remember God has a plan before he created the world; he knew that he would have his son walk among men.

5. WHY DID JESUS COME ONCE, AND WHY DID HE COME TO ISRAEL?

Now we understand why Jesus came when he came. Why didn't he come to Israel, rather than to America? Why didn't he come more than once?

Samuel was one of the great prophets of the day. God told Samuel to say this to David. *Samuel 7 12-14 NLV* **"'Furthermore, the Lord declares that he will make a house for you—a dynasty of kings! For when you die and are buried with your ancestors, I will raise up one of your descendants, your own offspring, and I will make his kingdom strong. He is the one who will build a house—a temple—for my name. And I will secure his royal throne forever. I will be his father, and he will be my son. If he sins, I will correct and discipline him with the rod, like any father would do.**

So it was God'S plan for Jesus to come through the line of David. God also promised Abraham's seed that he would be a blessing to many nations. God told Abraham to leave his land and go to a different land. The Lord had said to Abram, **"Leave your native country, your relatives, and your father's family, and go to the land that I will show you. I will make you into a great nation. I will bless you and make you famous, and you will be a blessing to others. I will bless those who bless you and curse those who treat you with contempt. All the families on earth will be blessed through you."**

God's plan was also for him to be born by a virgin. In *Isaiah 7 13-15* **13 Then Isaiah said, "Listen well, you royal family of David! Isn't it enough to exhaust human patience? Must you exhaust the patience of my God as well? 14 All right then, the Lord himself will give you the sign. Look! The virgin will conceive a child! She will give birth to a son and will call him Immanuel (which means 'God is with us').**

15 By the time this child is old enough to choose what is right and reject what is wrong, he will be eating yogurt and honey.

It was also God word that he wanted Jesus to be born in Israel and specifically Bethlehem In *Micah 2: 12-13* **"Someday, O Israel, I will gather you; I will gather the remnant who are left. I will bring you together again like sheep in a pen, like a flock in its pasture. Yes, your land will again be filled with noisy crowds 13 Your leader will break out and lead you out of exile, out through the gates of the enemy cities, back to your own land. Your king will lead you, the Lord himself will guide you."**

In *Micah 5 2-5* **But you, O Bethlehem Ephrathah, are only a small village among all the people of Judah. Yet a ruler of Israel, whose origins are in the distant past, will come from you on my behalf.3 The people of Israel will be abandoned to their enemies until the woman in labor gives birth. Then at last his fellow countrymen will return from exile to their own land.4 And he will stand to lead his flock with the Lord's strength, in the majesty of the name of the Lord his God. Then his people will live there undisturbed, for he will be highly honored around the world.5 And he will be the source of peace.**

So it was prophesied that Jesus would be born in Bethlehem and that he would be in Israel. That he would be in the line of David and the line of Abraham. That he would be born a virgin. So we know that God keeps his word. All the great prophets came through Israel from the original promise to Abraham. That he would be a blessing to many nations.

Why was Israel such a focus for God? Why not have some of the prophets in other parts of the world? WellGod, God is so knowledgeable. If there were other prophets in other parts of the world. The prophets would have got lost in translation. In the Inuit (Eskimo) language there are 50 words for snow. In Hebrew, there are a few words for snow. For example, there is no modern Hebrew word for sleet.

Some languages have multiple meanings for one word. In German, there is the word "bitte". In English, it means the following: 1. Pardon 2. Please 3. Go ahead 4. Here you go 5. Your welcome 6. Not at all. Also in German gut zu Vögeln sein means to be good to birds. gut zu vögeln sein means I want to have sex with you. Just changing the v from a capital to lowercase changes the whole meaning of the word.

The Old Testament was mostly written in Hebrew, and the New Testament was written in Greek. God has chosen one person to be a blessing to the whole nation. God didn't want confusion. He didn't want to show his prophets in other parts of the world. It would be harder to put the Bible together if there was a prophet in Africa and then another one in France. As I described before, language varies greatly from one place to another. The language just needed to be a couple of languages. Then from the Bible it could be translated into over 600 languages.

Why did Jesus come only once, why did he not come again? *1 Peter 21-23* **says For God called you to do good, even if it means suffering, just as Christ suffered for you. He is your example, and you must follow in his steps.22 He never sinned, nor ever deceived anyone.23 He did not retaliate when he was insulted, nor threaten revenge when he suffered. He left his case in the hands of God,who always judges fairly. Jesus had no flaws. Jesus died for our sins he couldn't improve on perfect. So that why he couldn't come back as a human. Jesus will come back, but that will happen during the end of days.**

God promised Abraham that the Jewish people would be a blessing to the whole world. All of these years later God is still blessing the Jewish people. The Jewish people continue to be a blessing to the whole world. Here is the list of some Jewish people who have major impacts in our world and have been blessings to the world.

Estee Lauder (Jewish) founder of (Estee Lauder Cosmetics over 11 billion dollars in Revenue)

Charles Revlon (Jewish) co founder of (Revlon (Cosmetics over 2.3 billion dollars in Revenue)

Louis Blaustein (Jewish) founder of (AMOCO (Energy) now BP 115 million dollars profit)

Leon Hess (Jewish) founder Ceo (Hess Energy 6.6 billion dollars in Revenue)

Michael Bloomberg (Jewish) former mayor of New York net worth 47 billion dollars

Mark Cuban (Jewish) owner of Dallas Mavericks worth net worth 3.3 billion dollars

Marcus Goldman (Jewish) co-owner of (Goldman Sachs net 37.7 billion dollars in revenue)

Samuel Sachs (Jewish) co-owner of (Goldman Sachs net worth 37.7 billion dollars in revenue)

David René de Rothschild (Jewish) current chairman of (Rothschild & Co with over 1 trillion dollars in revenue)

Burt Baskin (Jewish) founder of (Baskin Robbins revenue last year 828 million dollars)

Ben Cohen (Jewish) co founder of (Ben & Jerry's 132 million sales revenue)

Arnold Greenberg (Jewish) co founder of Snapple 5.97 million

Reuben Mattis (Jewish) founder of Haagen Dazs (ice cream) Nestle Parent company 89.46 billion dollars in revenue

Ziev Siegel (Jewish) co founder of Starbucks 19.6 billion dollars in revenue

Conclusion

You could see that Jewish people have been a blessing for the world. Some of the companies they have founded have been Starbucks, Haagen

Dazs, Baskin Robbins, Goldman Sachs. That was a God blessing to Abraham. That blessing still comes even today.

All of this was God'S plan. It was a God plan to focus on one area of the world that could be a blessing to the whole world. If there were a lot of people from many different areas, the message of the Bible would be less clear. Because language varies so much from one place to another, one word can have many different meanings.

We could not improve perfection. Jesus was perfect, and he died for sins. There was no reason for Jesus to come back if he did a perfect job. Jesus will come back, but he will come back as the spirit.

6. WHY IS THERE SUFFERING?

There are a couple of reasons why I think there is suffering and pain. First, if we have our hand over a fire. We get feelings of pain. If we didn't have pain, then your hand would burn up in a fire. So I think that God allows some pain to sometimes avoid more pain.

A. Also God uses pain as direction. You think about Candace Lightner, who founded Mothers Against Drunk Driving after her daughter was killed by a drunk driver. I could not imagine the pain that she must have felt from losing a child. Instead of being consumed with pain, she constructed her anger and her pain into something positive.

The Mother Against Drunk Driving (MADD) Organization helps create a minimum age of 21 for the drinking age. If the state did not enact the laws they would lose federal money for highways. In 2016, it was 70 billion dollars after the law was upheld by the Supreme Court. MADD has helped with laws regarding the blood alcohol limit.08 some states had.15 to be called legally drunk.

In 2002, MADD announced its "Eight-Point Plan." This consisted of:

1. Resuscitating the nation's efforts to prevent impaired driving.
2. Increasing driving while intoxicated (DWI)/driving under the influence (DUI) enforcement, especially the use of frequent, highly publicized sobriety checkpoints.
3. Enacting primary enforcement seat belt laws in all states.
4. Creating tougher, more comprehensive sanctions geared toward higher-risk drivers.
5. Developing a dedicated National Traffic Safety Fund.
6. Reducing underage drinking.
7. Increasing beer excise taxes to the same level as those for spirits.

8. Reinvigorating court monitoring programs.

Before MADD in the 1970's 60% of traffic fatalities involved alcohol. In 2008, that percentage was 32%. [9] They say 150,000 lives between 1982 and 2001 were saved because of MADD. All this was started by one kid dying tragically because of a drunk driver. One kid dies, but over 150,000 are saved.

B. (Why is there suffering?) Another reason for pain is that sometimes God wants to remind us that we need him, When there is a tragedy, what do news people say we need to do our prayers. Whether that would be a school shooting, a hurricane, a tornado, a terrorist attack and other tragedies. People start praying. There is a candlelight vigil. People realize pain overwhelms them, and they turn to God. They turn to God for answersGod. There was an increase in church attendance when the September 11 attacks happened. Whenever there is a tragedy, Church attendance never decreases. There is always an increase in attendance and prayers when there are these tragedies, I think God allows these things to happen to show how much we need God. God wants us to have a relationship with him.

C.(Why is there suffering?) What if I told you that suffering would be only for a couple of minutes of your life? That sadness would last a couple of minutes of your life. Any hopelessness would just last a couple of minutes. Joy and happiness would be the majority of your life. How would you view life? As humans, our vision is limited. We only see our years on earth. If we add our life in heaven, then suffering sadness only comes to.000000000000000000000000000000001 % of life.

As a metaphor, it would be like saying that the grass on the football field is horrible. When one of the blades on the football is dead. When you look at life as if this is just a small part of your life. We focus so much on what happens on this Earth that we don't focus on where we are going to spend most of our lives. Especially in today's society, we look so much for our immediate time rather than spending our time period. With that mindset, why should you sin against God. It may benefit you temporarily, but in the long run, it may plague you for the rest of your life. Like stealing you may get money now, but spend.9999999999999999999999999 paying for that mistake that you made. Is it worth it?

Another take on why there is suffering. If you look at any typical movie. There is an action star. Imagine Tom Cruise or Harrison Ford, there is some sort of obstacle. We get excitement from when the good guy triumphs over evil. Imagine watching a movie where we didn't see that. We just spent 90 minutes watching Tom Cruise. What would the movie be about? I can't think of any movie where someone had to overcome obstacles. In most movies, the Good guy would overcome the bad guy.

Imagine if your family were millionaires... Then you won the lottery. Would that be the same as if you were homeless and you would win the lottery. Imagine if nothing bad ever happened to you. Would heaven be the same, or would it be more special if there were pain? I would argue that pain is a spice in life. That makes life more special than if there wasn't any pain. Like if you were making a picture. You want to include many different colors, not just one. Life, you want more than Joy. You want all the spectrums of life: Joy, Pain, Sadness, Hope, Fear. This makes a beautiful picture of Life.

Just because there is pain does not mean that God gets pleasure from this. If you look at the scriptures, you see that God gets pleasure from things that bring life.

Conclusion

Pain is a useful tool to prevent damage. I gave an example of a hot stove. If we didn't have pain receptors, we would really hurt ourselves and have permanent damage. Sometimes when I have death or pain, it gives focus. We gave the example of MADD and how many people were saved because of someone's pain.

Pain gives life meaning. It gives life meaning because it gives us perspective. If we only have Joy then we would know pain. If everyday was sunny, would we enjoy sunny days as much as if we had both sunny and rainy days? Also when we have a tragedy, it can bring people together. Pain has purpose

We also have to keep in mind that we are only on Earth for a blink of an eye. Eternity is such a long time. We only have pain for such a short period of time. I gave that our life is a blade of grass on a football field.

7. WHY DO INNOCENT PEOPLE DIE?

It seems like things are more tragic when people die who weren't supposed to die. Like when a drunk driver kills another family in another car. Or when a drug deal goes bad, and an innocent child gets killed. I don't have all the answers, but I attempt to answer them.

A.What if I told you that you don't have to work, anymore, that there is no hunger, no more bills? No worries anymore, the secrets of life revealed. Well, all this would happen when you get to heaven. So why not Godhave God take his best into heaven. That is going to be the innocent child that is going to be the innocent child. Rather than have a lifetime of paying bills with the evils of this world, they get to skip all that and go to heaven. There are popular movies and stories that talk about heaven. There is nobody who says heaven was horrible or just ok. Everyone is fundamentally changed into a different person. So why not God to take innocence?

Don't forget that Jesus Christ was killed. *1 Peter 1:19* **It was the precious blood of Christ, the sinless, spotless Lamb of God**. *Luke 1:35* **The angel answered and said to her, "The Holy Spirit will come upon you, and the power of the Most High will overshadow you; and for that reason the holy Child shall be called the Son of God. So Christ was innocent of them all, so Christ was whipped and put a crown of thorns are there.** In *Matthew 27: 26 29* **And after twisting together a crown of thorns, they put it on His head, and put a reed in His right hand [as a scepter]. Kneeling before Him, they ridiculed Him, saying, "Hail (rejoice), King of the Jews!" 30 They spat on Him, and took the reed and struck Him repeatedly on the head. 31 After they finished ridiculing Him, they stripped Him of the scarlet robe and put His own clothes on Him, and led Him away to crucify Him. So if God allows that to happen to his only son, and he takes him at the young age of 33. Then why not take Little Susie or someone else?**

B. (Why do innocent people die?) When we question God we are thinking we know more than God. That is the original sin. When Adam, the first man, took the apple, he was saying that Eve and the serpent knew more than God. When we say that this child should be killed you are saying you know better than God. Sometimes God has some intended benefit from someone dying.

Imagine if you were a follower of Jesus, and you watched Jesus heal the sick. You watched hundreds hear Jesus' word. You watched Jesus be compassionate to people. This same person you see gets beaten. You see them with a crown of thorns. Then you watch the Son of God die. If you look at the smaller picture, this would be horrible. Looking at the bigger picture, Jesus died for our sins. Jesus died for our sins. *Romans 5:8* **But God showed his great love for us by sending Christ to die for us while we were still sinners. This is all part of God plan** *Zechariah 12:10 NLV* **"I will pour out on the house of David and on the people of Jerusalem, the Spirit of grace (unmerited favor) and supplication. And they will look at Me whom they have pierced; and they will mourn for Him as one mourns for an only son, and they will weep bitterly over Him as one who weeps bitterly over a firstborn.**

If you remember earlier in this book. Then Christine Lightner's daughter was killed by a drunk driver. From that tragedy, MAD was created. If you remember, MAD cracked down on drunk driving. MAD has been attributed to saving over 150,000. They also cut the alcohol death percentage from 60 % before MAD to where it is closer to 30% now. Imagine if God didn't allow Christine Lightner's daughter to die.

From that one death many people were saved because of that. Who can say but your loved one in some way could make. Or that in some way someone's death could inspire someone to do great. When we look at great men a lot of the time they have died before they could. Abraham Lincoln died before he could see the slaves have true freedom. We had Martin Luther King Jr who was killed before he could see equal rights for black people.

C.(Why do innocent people die(suffer)? What about bad people, why don't they just die? Don't forget what happens when we die. If we don't repent we go to hell. We are permanently separated from God. Could you watch one of your children getting burned in the lake of fire. God feels that we are his children.

In *2 Peter 3:9 MSG* **Don't overlook the obvious here, friends. With God, one day is as good as a thousand years, a thousand years as a day. God isn't late with his promise as some measure lateness. He is restraining himself on account of you, holding back the End because he doesn't want anyone lost. He's giving everyone space and time to change. God doesn't want to judge us because he wants us to have a relationship with us he wants us to get us to know him.**

John 3:16 -21 NLV **For this is how God loved the world: He gave[g] his one and only Son, so that everyone who believes in him will not perish but have eternal life. 17 God sent his Son into the world not to judge the world, but to save the world through him.18 "There is no judgment against anyone who believes in him. But anyone who does not believe in him has already been judged for not believing in God's one and only Son. 19 And the judgment is based on this fact: God's light came into the world, but people loved the darkness more than the light, for their actions were evil. 20 All who do evil hate the light and refuse to go near it for fear their sins will be exposed. 21 But those who do what is right come to the light so others can see that they are doing what God wants.**

You put a child on the Earth. While he was there, he did some awesome things. You felt really proud *Matthew 3 :16* says [16] **After his baptism, as Jesus came up out of the water, the heavens were opened[and he saw the Spirit of God descending like a dove and settling on him.** [17] **And a voice from heaven said, "This is my dearly loved Son, who brings me great joy." These same people you watch beat up your son and kill them. And you do this so that it would be easier for so that people would have eternal life.**

You wouldn't go through all those steps for nothing. It shows how much he wants us to repent and change our ways. Imagine if we did something wrong and were instantly killed. We would all die when we were 3 -4, and if we did something wrong, then we would die. There wouldn't be anybody alive. There wouldn't be anybody alive. That is why he lets us make mistakes over and over, but a chance to repent.

Conclusion

We think that someone who dies is a bad thing. What if it's not always a bad thing?

We have to remember that Jesus was without sin, and he was killed at 33 years old. If he could be killed at any time, why not anyone else be killed? Also, remember that death sometimes gives us focus. Remember my example of the MADD with the death of Christine Lightner's daughter, they were able to save a lot more lives. From that one death, it is estimated to save over 150,000 lives.

Let's not forget that Jesus Christ died for humanity. He died a sinner. Sometimes death has a purpose that we may not know about. Sometimes death could start a war like Crispus Attucks, which led to the American Revolution. Or the death of Abraham Lincoln pulled the country together. If a clock represents our existence. Our lives on earth would be a second of the day. Most of our existence would be in heaven. We can't focus on the second that we are on Earth or why the innocence may die. We need to focus on heaven and get everyone else to heaven. This is where the majority of our existence. We shouldn't forget that.

8. I UNDERSTAND WHY SINNERS DON'T DIE, BUT WHAT ABOUT MURDERERS? WHY DO THEY LIVE?

God views any sin as bad. What is sin? Sin is separation from God. *Romans 6:23* [23] **For the wages of sin is death, but the free gift of God is eternal life through Christ Jesus our Lord. So the way God views it, anything that isn't God is sin. So whether you say something mean or kill someone, it is still bad. Now, in the eyes of God the worst is the Ten Commandments, which are the following**

1. **You shall have no other gods before Me.**
2. **You shall not make idols.**
3. **You shall not take the name of the LORD your God in vain.**
4. **Remember the Sabbath day, to keep it holy.**
5. **Honor your father and your mother.**
6. **You shall not murder.**
7. **You shall not commit adultery.**
8. **You shall not steal.**
9. **You shall not bear false witness against your neighbor.**
10. **You shall not covet.**

It's not that the things that we do are so bad and we don't deserve death. We all deserve death for things that we do, It's because God is good that he gives us years of chances to repent. *Exodus 2 :11-12* [11] **Many years later, when Moses had grown up, he went out to visit his own people, the Hebrews, and he saw how hard they were forced to work. During his visit, he saw an Egyptian beating one of his fellow Hebrews.** [12] **After looking in all directions to make sure no one was watching,**

Moses killed the Egyptian and hid the body in the sand. In *Exodus 33 15 -23 NLV* [15] **Then Moses said, "If you don't personally go with us, don't make us leave this place. 16 How will anyone know that you look favorably on me—on me and on your people—if you don't go with us? For your presence among us sets your people and me apart from all other people on the earth."[17] The Lord replied to Moses, "I will indeed do what you have asked, for I look favorably on you, and I know you by name."[18] Moses responded, "Then show me your glorious presence."[19] The Lord replied, "I will make all my goodness pass before you, and I will call out my name, Yahweh, before you. For I will show mercy to anyone I choose, and I will show compassion to anyone I choose. [20] But you may not look directly at my face, for no one may see me and live."[21] The Lord continued, "Look, stand near me on this rock. [22] As my glorious presence passes by, I will hide you in the crevice of the rock and cover you with my hand until I have passed by. [23] Then I will remove my hand and let you see me from behind. But my face will not be seen."**

So this is somebody who has killed someone and buried him in the sands. Moses overcame his sin. Clearly, killing someone is one of the Ten Commandments. If God judged Moses right when he killed the Egyptian in the sands, Moses was able to change that to the point that God states to him I have favor, and you know my name, Moses was able to see God presence. Few people have seen his presence. It is something special to see his presence. Of the people Godwho saw God, most would have seen God in a dream. When Moses saw his God he couldn't see him directly; he saw his presence after he walked past a rock. It shows all magnificient God is, and it also shows how favored Moses is. So if Moses can come from a man who killed someone to being a great man. Then who's to say who the next Moses is?

Conclusion

We have all fallen short of the glory of God. God gives us the opportunity to turn our lives to him. I gave the example of Moses, who killed someone and buried them in the sands. So if God killed Moses on the spot after he killed the Egyptian, then we wouldn't have the book of Moses, Genesis, Exodus, Leviticus, Numbers, and Deuteronomy.

Or in the New Testament, we had Paul, who was part Pharisee. The Pharisees were a religious sect of Judaism that would persecute the Christians. They were instrumental in killing Jesus. He went from hating Christians to becoming the biggest writer of the New Testament. Paul wrote 13 of the 27 books of the New Testament. God gives us all the opportunity to change our ways. Some of the most evil people have become the greatest people in life.

9. WHY DON'T WE SEE MIRACLES, OR WHY DON'T I SEE THE WONDERS OF GOD?

If there are miracles all through the Bible. In the Bible, it says God doesn't change, so why don't we see miracles?

A.There could be a couple of reasons why miracles are not happening for the first thing in order for there to be a miracle, you would have to believe it. If you don't believe miracles could happen, then it won't happen. That was the case in the Bible and that is the case in real life. As people move away from their faith, there will be fewer miracles. If you look at churches where the congressman doesn't believe in miracles, then it won't happen. Then you have a church where they do believe in miracles, then that is what you are going to see.

B.Why don't we see miracles, or why don't I see the wonders of God? Sin is the separation for God. When you are separate from God. God can intervene. God can't really intervene when there is that separation. Look through the Bible anytime there was a miracle, there was a closeness from God. You have to hear God's word, obey his word, and seek him at all times. As each generation comes and goes, we get further and further away from him. It started with Adam and has gone downhill from there.

C. Why don't we see miracles, or why don't I see the wonders of God? With today's use of televisions, phones, apps, and today's modern society. Everybody is distracted from God. So nobody has quiet time with God. Before our modern times, there was time when we had to have a conversation with God. With our faces on our screens, we can’t see all the miracles around us. With our modern society, we feel like we don't need God, we have Doctors. If we have a headache, we pop a pill for that, if a kid misbehaves, we pop a bill for that, if we have depression, we pop a pill for that. I am not saying we shouldn't have pills, but we should also seek God. If we don't ask for a miracle, we can't have it. People are quicker to believe in Men than in God

In societies where it is agarin societies. There is a stronger reliance on God. You believed in God for your crops. You believed in God for healing; there were not always doctors available. You believe in God's protection. You could be attacked by wild animals. Today, there is greater separation from God. We live more and more in a society that focuses on screens rather than the sky.

God is still doing miracles every day. I would argue that there are more Miracles now than there were in Jesus' time. God is healing people's sight, bringing people from death, God is giving visons, God is providing Children when doctors say women are barron. However, when miracles happen, some are quicker to explain things away.

Conclusion

There are several reasons why we don't see miracles. As a modern society, we don't feel like we need him. We were born separated from him. So since we were born separated from him, we don't have a relationship with him. A lot of times, we need to have a relationship in order to get prayers done.

We aren't doing God's business. We are only thinking of ourselves, and we are not thinking of other people. God will always equip us to do the business of God. When God wanted something to get done, he got it done through miracles.

Miracles are happening every day. However, as a modern society. We stopped believing in God and started believing in us. But one day every Phliphians 2:10 every knee show bow and every Tongue shall confess.

10. WHY DON'T MY PRAYERS GET ANSWERED?

That popular question people ask: why doesn't this happen? I will attempt to answer in several ways.

A. The prayers aren't what is best for you. Like, if you're robbing someone and you pray not to get caught, that's not what we pray for when we pray to help with sin. God will never help us fulfill sin. Now he will get out of sin. But our hearts must be pure. A lot of times, our hearts are not pure, and so he will not get us out of the mess. Sometimes you may feel something is good for you, but it isn't. God please give me tickets for a concert. God knows a whole lot more, and maybe this concert is a bit better for you. When we question God we say we know better than God. So, going to the concert may not be a sin, but there may be something that you could make better use of your time.

In James 5 16 NLV " **The earnest prayer of a righteous person has great power and produces wonderful results."** 1 Peter 3- 12 "**The eyes of the Lord watch over those who do right, and his ears are open to their prayers."**

B.Why don't my prayers get answered? God does not answer prayers like magic tricks. Matthew 4 5-7 NLV **Then the devil took Him up into the holy city, set Him on the pinnacle of the temple, [6] and said to Him, "If You are the Son of God, throw Yourself down. For it is written: ' He shall give His angels charge over you,' and, 'In their hands they shall bear you up, Lest you dash your foot against a stone.'" [7] Jesus said to him, "It is written again, 'You shall not tempt the Lord your God.'" So God is not a parlor trick. Hey God move this cup from one side of the room.**

If you look at when prayers get answered in the Bible. They get answered most when you are doing God's work. When you are doing God's work, then God will answer prayers to do God's work. So the converse of that is that if you're not doing God's work, then your prayers might not be answered. What might you say is God's workGod? God's work is telling people about.

Jesus understood early on that his purpose was doing his father's business. *Luke 2 41-50* **41 Every year Jesus' parents went to Jerusalem for the Passover festival. 42 When Jesus was twelve years old, they attended the festival as usual. 43 After the celebration was over, they started home to Nazareth, but Jesus stayed behind in Jerusalem. His parents didn't miss him at first, 44 because they assumed he was among the other travelers. But when he didn't show up that evening, they started looking for him among their relatives and friends.**

45 When they couldn't find him, they went back to Jerusalem to search for him there. 46 Three days later, they finally discovered him in the Temple, sitting among the religious teachers, listening to them and asking questions. 47 All who heard him were amazed at his understanding and his answers.

48 His parents didn't know what to think. "Son," his mother said to him, "why have you done this to us? Your father and I have been frantic, searching for you everywhere."

49 "But why did you need to search?" he asked. "Didn't you know that I must be in my Father's house?" 50 But they didn't understand what he meant.

In *Mark 16 15-18* **And then he told them, "Go into all the world and preach the Good News to everyone. 16 Anyone who believes and is baptized will be saved. But anyone who refuses to believe will be condemned. 17 These miraculous signs will accompany those who believe: They will cast out demons in my name, and they will speak in new languages. 18 They will be able to handle snakes with safety, and if they drink anything poisonous, it won't hurt them. They will be able to place their hands on the sick, and they will be healed."**

This verse is clear: if we need to believe in order for our prayers to be answered, we need to be spreading the good news. Not always, but prayers are help from God in getting his will done on earth. What is the good news? The good news is that we have eternal salvation if we believe in Jesus Christ. While doing that, we will see miraculous signs.

Conclusion

When we are doing God's business, we can do all things. Some people say that when we prayGod, we align our prayers with God's will through our words. So when we pray, and we are selfish God may not answer the prayer. If the prayer is committing a sin, God will not answer that

God is about doing God's business. Prayers aren't for our selfish needs. God will provide you food shelter, and everything you need. You may not get everything you want, but you will get what you need.

God is more likely to listen if you're righteous. *Proverbs 15 :8* **The Lord detests the sacrifice of the wicked but he delights in the prayers of the upright.** So if we are righteous doing God work and we are not doing selfish things. You're most likely to get your prayer answered.

11. WHAT IS THE TRINITY?

The trinity is the 3 pieces of God. The Father, the son and the holy spirit. Some people can think how there could be three parts of God. Is God still God? God is still God. Important thing to remember if the 3 pieces of God were there in the beginning. In *Genesis 1:26,* **it states Then God said, "Let us make human beings in our image, to be like us. They will reign over the fish in the sea, the birds in the sky, the livestock, all the wild animals on the earth, and the small animals that scurry along the ground."From the text, it shows there were more than one piece of God there.**

We got to understand that God is not bound by time. God is in the past, present, and future. God saw that man would need Jesus ahead of time. So when the earth was formed he was there. He knew that he would need Jesus. And he knew that eventually there would be the end of man.

For some people have a hard time with the concept of three things being one. So I looked at three other important things. Matter is in 3 different states. Solid, liquid, and gas. So water is the same as ice, yet different than ice. Water vapor is the same as water, yet different. They are just a different form.

When you measure anything it is length, width and height. Length is like width, width is like height. I think it's no mistake that the smallest-sided shape is a triangle. There is a part of math devoted to the trigonometry of a three-sided figure. 3 is a sacred number.

If we look at the example of the water. God would be the ice. Jesus would be water, and the holy spirit would be the water vapor. In Matthew 28 :18 **18 Jesus came and told his disciples, "I have been given all authority in heaven and on earth. 19 Therefore, go and make disciples of all the nations, baptizing them in the name of the Father and the Son and the Holy Spirit.**

2 Corinthians 14 [14]**The amazing grace of the Master, Jesus Christ, the extravagant love of God, the intimate friendship of the Holy Spirit, be with all of you.**

How would you recognize the difference between the Holy Spirit, the Father, and Jesus Christ? Well, the Holy Spirit is in the dwelling of God's spirit. That is why it is said to be the water vapor in the example of the water example. This is your conscience. It's the inner voice that tells you what to do.

Jesus is the bridge from man to God. *John 14:6-7* [6]**Jesus told him, "I am the way, the truth, and the life. No one can come to the Father except through me.** [7] **If you had really known me, you would know who my Father is. From now on, you do know him and have seen him!." Jesus would be considered the water in the water ice and water vapor example. Jesus is the second Adam. He starts a new relationship with God the father. He reconciles man with God.**

The Father God he is the ice in the example of water ice and vapor. The father God is the head of the the tri-God. *John 6: 38-39 NLV* **"For I have come down from heaven, not to do My own will, but the will of Him who sent Me. "This is the will of Him who sent Me, that of all that He has given Me I lose nothing, but raise it up on the last day. The Father God would be the word that manifests into something real.**

Conclusion

The trinity is one of the hardest things to describe in Christian faith. There's something sacred about three. Three is the smallest sides for a shape a triangle. It is described for the measurement of width, length, and depth. 3 is the basic level of matter: solid, liquid, and gas. You could see how 3 is such an important number.

Water, when frozen, is ice; when liquid, it is water; and when vaporized, it is vapor. They are all the same thing, yet they are all different. All these are different but yet the same. That's the way God is. He is the Holy Spirit, Jesus, and God.

The Holy Spirit is the little voice that guides us. Jesus is the bridge between God and man. God is the architect of everything.

12. WHY DO WE PRAY?

And why is it important to pray out loud? John 1:1-5 In the beginning the Word already existed. The Word was with God, and the Word was God.[2] He existed in the beginning with God.[3] God created everything through him, and nothing was created except through him.[4] The Word gave life to everything that was created, and his life brought light to everyone.[5] The light shines in the darkness, and the darkness can never extinguish it. So God created the world with words.

So, in order for there to be action, there has to be words. To demonstrate how important words are, I wanted to share an experiment. A guy had three plants. One plant he said hateful words. Hateful words like I hate you, you're ugly you're a terrible plant and other words. One plant he said nothing to. The last plant he said, positive things. Positive words like your beautiful, your a great plant and other positive words. The plant that he said hateful words did not grow it withered and died. The second plant that he stated nothing there was some growth. The last one that he positive there was much growth that the one they said nothing. There are more experiments out there. There are experiments with rice cereal and other experiments. All lead that when words are negative there is death or slow growth. When nothing is said there is slow growth or no growth. When talking positive words that showed the most growth.

What does this mean? Bottom line: words matter. They did at the start of the universe and they still do today. Ok, now that we establish the importance of words why do we pray. Prayer is a petition from God to intercede. Jesus needed it in order to accomplish what he needed to do. If Jesus needed prayer, then we certainly needed it. There was something big moments when Jesus needed prayer. Who better to show good examples when you would want to use prayer?

A. Baptism - Jesus said to pray to God *Luke 3 21-22*[21] **One day when the crowds were being baptized, Jesus himself was baptized. As he was praying, the heavens opened,** [22] **and the Holy Spirit, in bodily form, descended on him like a dove. And a voice from heaven said, "You are my dearly loved Son, and you bring me great joy. So from this you would want to pray for big moments in your life**. So weddings graduations baptisms you want God to be with you in those moments. Because in these moments you want important people in your life there for these moments you are showing God that he is important.

B. Jesus prayed for selecting the 12 disciples *Luke 6* **12 One day soon afterward Jesus went up on a mountain to pray, and he prayed to God all night. 13 At daybreak he called together all of his disciples and chose twelve of them to be apostles. So anytime we are making a big decision we want to use God.** Remember God is infinitely smarter than you are. He wants us to use him. Imagine if Adam had called on God should I eat this fruit of knowledge. We would not have sinned. So if we are making any important decision buying a car or a house. It's important to ask for God discernment.

C. Jesus predicts his death Luke 9 -21 **18 One day Jesus left the crowds to pray alone. Only his disciples were with him, and he asked them, "Who do people say I am?"19 "Well," they replied, "some say John the Baptist, some say Elijah, and others say you are one of the other ancient prophets risen from the dead."20 Then he asked them, "But who do you say I am?" Peter replied, "You are the Messiah[sent from God!"21 Jesus warned his disciples not to tell anyone who he was. 22 "The Son of Man[e] must suffer many terrible things," he said. "He will be rejected by the elders, the leading priests, and the teachers of religious law. He will be killed, but on the third day he will be raised from the dead."** So if constant prayer it is conversation with God. God is the same as it was 2000 years ago and he is the same today. If you look at the people throughout the Bible. They were able to tell what going to happen in the future. There is some debate but there are said to be

4 major prophets of the Bible and 12 minor prophets of the Bible. Sometimes God could tell you something is going to happen so you could change it, and sometimes he tells you something so you could prepare for it. In Jesus case, he was told so he could prepare for it.

D. Revelation *Luke 9 - 28- 30* **28 About eight days later Jesus took Peter, John, and James up on a mountain to pray. 29 And as he was praying, the appearance of his face was transformed, and his clothes became dazzling white. 30 Suddenly, two men, Moses and Elijah, appeared and began talking with Jesus.31 They were glorious to see. And they were speaking about his exodus from this world, which was about to be fulfilled in Jerusalem. So Jesus prayed for revelation of God. So this could work for helping someone else to believe in God. You could pray for a revelation from God. As what Jesus did for his disciples. His disciples had already seen miracles but this show we could all use more believe to strengthen our believes.**

E. Miracles *John 11 37-43* **37 But some said, "This man healed a blind man. Couldn't he have kept Lazarus from dying?"38 Jesus was still angry as he arrived at the tomb, a cave with a stone rolled across its entrance. 39 "Roll the stone aside," Jesus told them. But Martha, the dead man's sister, protested, "Lord, he has been dead for four days. The smell will be terrible."40 Jesus responded, "Didn't I tell you that you would see God's glory if you believe?" 41 So they rolled the stone aside. Then Jesus looked up to heaven and said, "Father, thank you for hearing me. 42 You always hear me, but I said it out loud for the sake of all these people standing here, so that they will believe you sent me." 43 Then Jesus shouted, "Lazarus, come out!" 44 And the dead man came out, his hands and feet bound in grave clothes, his face wrapped in a head cloth. Jesus told them, "Unwrap him and let him go!"** So clearly, Jesus was in communication with God the Father. So expect the unexpected with God. Normal rules don't apply. You have to pray for it and have faith that it is done. *Proverbs 3 5-6* **Trust in the Lord with all your heart;do not depend on your own**

understanding.6 Seek his will in all you do, and he will show you which path to take. So pray in miracles and expect it. Jesus expected Lazarus to be raised from the death that is how we should pray like.

F. Comfort/strength In *Luke 22: 39- 44* **39 Then, accompanied by the disciples, Jesus left the upstairs room and went as usual to the Mount of Olives. 40 There he told them, "Pray that you will not give in to temptation."41 He walked away, about a stone's throw, and knelt down and prayed,42 "Father, if you are willing, please take this cup of suffering away from me. Yet I want your will to be done, not mine." 43 Then an angel from heaven appeared and strengthened him. 44 He prayed more fervently, and he was in such agony of spirit that his sweat fell to the ground like great drops of blood**. You could see that Jesus the son of God needed strength. So God has made us that in some circumstances it will be overwhelming to us so we would need the help of God to overcome it. Jesus prayed so hard that we started to bleed, which shows you how hard he was praying.

These are by no means a complete list of all the reasons you should pray. You should always be thankful for what you have. The next question is I will ask is how you should pray. But I wanted to layout the reasons Jesus prayed,

Conclusion

The world was created by words. Words are very powerful. In an experiment with plants. The guy spoke words pleasant words and the plants grew the most. When the plants were told bad words, they did not grow.

Jesus used prayer when he was deciding his 12 disciples. So I think whenever we are going to make a big decision, we should be like Jesus and ask our heavenly father for advice. Jesus also asked God about what going to happen. Remember, God is not bound by time. He is the Alpha and the Omega; the start and the end. When Jesus did miracles, he asked for God help. If we want something that is not humanly possible, we must ask for God help.

When Jesus was about to be killed. He prayed for God help for strength. So when we are struggling with anything, we should ask for God's help. That could be strength, that could be wisdom, that could insight, and anything else that you may need. Remember God is unlimited and unbound.

13. HOW TO PRAY?

I think there is no more important teacher than Jesus. In *Matthew 6:5* NLT, **"When you pray, don't be like the hypocrites who love to pray publicly on street corners and in the synagogues where everyone can see them. I tell you the truth, that is all the reward they will ever get. 6 But when you pray, go away by yourself, shut the door behind you, and pray to your Father in private. Then your Father, who sees everything, will reward you.7 "When you pray, don't babble on and on as the Gentiles do. They think their prayers are answered merely by repeating**

Jesus actually tells us how to pray, Matthew 6 9-13 NLV **Our Father in heaven, may your name be kept holy.** So the first line you want to start off by saying Godis addressing him as our father. The second thing he wants to keep God's name holy. So you don't want to curse him or say why didn't you do this, **10 May your Kingdom come soon. May your will be done on earth, as it is in heaven**. So when you pray you want to pray that God will be done. God will be that everyone will have a relationship with himGod; God will be, if we would keep his commandments. God will would be that we would serve him instead of God serving us. God will would be there would be no wars. **11 Give us today the food we need,12 and forgive us our sins, as we have forgiven those who sin against us** **13 And don't let us yield to temptation, but rescue us from the evil one.** So we want to be thankful for the food we receive. Also, we should continue to pray for God to provide everything that we need in daily life. In the original King James version, **"give us our daily bread**." "Daily bread" is what we would need in daily life. So that would be home, a job, food, transportation needs, and common products to live. We must forgive the people who have done us wrong. So we must pray for this help. We all have sinned, so we must forgive them. When Jesus was whipped and crucified, he asked for them to be given, We also should pray to get away from temptations. Jesus Christ was tempted, so no matter how good you are. We want to pray for God to lead us from temptation. God help us do better.

For Yours is the kingdom and the power and the glory forever. Amen. The last line says God this is your Kingdom. So when we pray you want to pray and let God know they have the power and glory. You want pray that God kingdom continues.

So lets says let's give an example of prayer using what Jesus Christ example that he gave us. Let's say there is a hurricane coming and you want to pray for safety, so you would start.

Father God, You are so HOLY and GOOD. Everyday we want to have your Kingdom to be like on earth has. Father God we have met up to your GLORY, Father. We have sinned and have not done what you wanted done. Father God help us do better. Forgive those who have sinned against us. Father God, we want to thank you for providing us with food, shelter, and a good job. We ask your divine blessing to prevent harm from the hurricane. We thank you God

their words again and again. **8** Don't be like them, for your Father knows exactly what you need even before you ask him.

When you pray as what Jesus told us. You don't want to do it just to do it. You want to do it in private. In *John 1 :1-4* **In the beginning the Word already existed. The Word was with God and the Word was God.[2] He existed in the beginning with God. [3] God created everything through him, and nothing was created except through him.4 The Word gave life to everything that was created, and his life brought light to everyone. Prayer is more effective when you words are outloud. Remember that is how the earth was created.**

Conclusion

People may not realize this, but Jesus tells us how to pray. Who better to tell us how to pray better than Jesus. Who would know better God better than his only son. The first thing that he tells us is to let God know that he is holy. The next thing is that you want to align with the father. That's when you pray,, you want to pray something that aligns with God's thoughts.

In the Lord's prayer, Jesus also states that he wants us to be thankful for what we have. Jesus tells us that we need to forgive our sins as we forgive those who have sinned against us. We should also ask God to lead us from temptation.

You want to pray that the kingdom be on earth as it is in heaven. What that means is a set of rules. That's what a Kingdom is. It starts with the king. If the King says jump, you say how high. In a kingdom, there is hierarchy when it comes to who rules whom. So, Prince (Jesus) is second in command. We must follow the laws that are in place. On earth as it is in heaven. That line also means we need to expand God's territory. As member of this Kingdom, we must expand our territory by reminding people who is king. And what the rules for this kingdom is (Bible).

14. HOW DO YOU KNOW YOU ARE A GOOD CHRISTIAN?

I am sure there are many opinions for what makes a good Christian. Here I have laid out some things that make a good Christian. You may agree with some of them and disagree with others. I tried to match it up with what the Bible teaches.

1. Do you follow Christian principles?
2. Do you know what the Bible says? Are you reading the Bible on a regular basis.
3. Do you have a good understanding of what the Bible says? Do you know the majors profits of the Bible do you know who the disciples of Jesus was
4. Do you follow Christ more than you follow the world? If your friends were doing something against what the Bible teaches. Would you do your friends told or would you do what the Bible tells you?
5. Do you feel the presence of God? Do you feel the Holy Spirit when praying, worshiping, or speaking the word? Do you cry or have emotional breakdowns when praying?
6. Has miracles happened in your life? Have you see anything that can't be answered scientifically? Have you had prayers answered or have you been in presence of where someone else's prayers was answered.
7. Have you been able to see a situation in God's eyes? Have you seen a different perspective than how the world would see a situation. Do you have an insight into a situation without having a different insight gift.
8. Have you been able to have "Gifts of the Spirit" *1 Corinthians 12 8-10* **these gifts are signs of that you a Christian. Nobody has all the gifts and some gifts are stronger than other gifts,. In order**

for you to get the gifts, you must pray that you have these gifts or increase them.

a. **Prophecy** able to see the things that will happen in the future and being able to prepare things that will happen in the future. Have you been able to tell what was going to happen in the future? Have you had any dreams of what was going to happen? Have an six sense that something bad was going to happen. Have you had an inner voice tell you what going to happen?

b. **Evangelist** There is a gift of being able to speak to other people. Being able to speak about Jesus. Have you been able to strike up a conversation with a stranger? Are you able to share your faith on a regular basis? Are you able to go to door to door sharing the faith?

c. **Pastor** being a leader in the church, being the head of the church. Are you able to teach others about the faith? Are you charismatic? Are you able to delegate responsibilities. Are you passionate about Jesus? Do you have good moral character?

d. **Teaching** (Sunday school teacher), someone who is able to teach the Christian faith? Are you able to teach the faith? Are you knowledgeable about the Bible. Do you show patience. Are you able to engage people? Do you like kids, and are you creative? Are you able to bring the Bible alive in different ways?

e. **Service** The gift of being able to serve other people. Are you faithful to your church? Do you show up to volunteer on a regular basis? Are you ok serving without getting something

f. **Giving** the ability to give. The ability to give without looking for something in return. The ability to give old toys, clothes, and small appliances, as well as time and money.

g. **Words of Wisdom** - The quality of having experience knowledge, and good judgement. Are you able to have supernatural ability to hear from God. Where someone else would say, "Don't do that," or

"Does not make sense." You get a word of God which will give you some wisdom. Wisdom is the ability to look beyond the present. Solomon was said to be the wisest man of his time. Solomon is where we get our proverbs. Solomon said that he gets his wisdom from God.

h. **Words of Knowledge**- Facts or information facts, information, and skills acquired by a person through experience or education. Whereas words of wisdom is similar gift to words of Knowledge. Words of knowledge has to do more with the present. Have you ever evangelized to someone and sometimes God will tell you something that going wrong with the person that you are evangelizing with. Have you ever taken a test and couldn't remember what the answer was? When you called on God and could remember it.

i. **Faith** - complete trust in God. Trust God and lead not to your own understanding. As Christian followers, we must have faith, but this is supernatural faith. I think of the centurion soldier in *Matthew 8 5-13 NLV* **5 When Jesus had entered Capernaum, a centurion came to him, asking for help. 6 "Lord," he said, "my servant lies at home paralyzed, suffering terribly."7 Jesus said to him, "Shall I come and heal him?"8 The centurion replied, "Lord, I do not deserve to have you come under my roof. But just say the word, and my servant will be healed. 9 For I myself am a man under authority, with soldiers under me. I tell this one, 'Go,' and he goes; and that one, 'Come,' and he comes. I say to my servant, 'Do this,' and he does it."10 When Jesus heard this, he was amazed and said to those following him, "Truly I tell you, I have not found anyone in Israel with such great faith. 11 I say to you that many will come from the east and the west, and will take their places at the feast with Abraham, Isaac and Jacob in the kingdom of heaven. 12 But the subjects of the kingdom will be thrown outside, into the darkness, where there will be weeping and gnashing of teeth."13 Then Jesus**

said to the centurion, "Go! Let it be done just as you believed it would." And his servant was healed at that moment. So, where were the other people that Jesus had to present because of the Romans? Jesus said he had more faith than anybody else. This Roman soldier believed in the person being healed, even though he could not see it or feel it. That's what the super faith is beyond what a typical Christian would believe. Whereas people would have see it or feel it this person of super faith would not.

j. **Gifts of healing.** The ability to heal someone. This is the ability to heal someone by laying on of hands, putting anointed oil on, or just praying. Sometimes the healing would be right away, and sometimes it would be delayed. It's important to remember it's not you healing, you are the vessel for someone else to be healed.

k. **Discerning of spirits**: This is an interpretation of what good spirits and bad spirits are. This could manifest in that you have a "funny feeling about someone." So you take precautions about with someone. Sometimes you may have a good feeling about someone. So, this person, you may come to know them. It also comes to understanding the Bible better. You may also understand spiritual things, such as dreams or things you may not understand. There was great prophets of the Bible that were able to interpret dreams.

l. **Tongues** the ability to talk in an unknown language. It's the ability to talk in different languages in the Bible. In *Acts 2:* 1-12 **On the day of Pentecost[a] all the believers were meeting together in one place. 2 Suddenly, there was a sound from heaven like the roaring of a mighty windstorm, and it filled the house where they were sitting.3 Then, what looked like flames or tongues of fire appeared and settled on each of them. 4 And everyone present was filled with the Holy Spirit and began speaking in other languages,[b] as the Holy Spirit gave them this ability.5 At that time, there were devout Jews from every nation living in Jerusalem. 6 When they heard the loud noise, everyone came**

running, and they were bewildered to hear their own languages being spoken by the believers.7 They were completely amazed. "How can this be?" they exclaimed. "These people are all from Galilee, 8 and yet we hear them speaking in our own native languages! 9 Here we are—Parthians, Medes, Elamites, people from Mesopotamia, Judea, Cappadocia, Pontus, the province of Asia, 10 Phrygia, Pamphylia, Egypt, and the areas of Libya around Cyrene, visitors from Rome 11 (both Jews and converts to Judaism), Cretans, and Arabs. And we all hear these people speaking in our own languages about the wonderful things God has done!" 12 They stood there amazed and perplexed. "What can this mean?" they asked each other. So there have been stories of modern day where people would evangelize with other people. People have been known to speak the language of the unbeliever. God does this so the unbeliever could believe. Imagine if someone spoke your language. That would be a confirmation there is a God. There was another verse in the Bible: "Let love be your highest goal!" But you should also desire the special abilities the Spirit gives—especially the ability to prophesy.*1 Corinthians 2-5* **2 For if you have the ability to speak in tongues,] you will be talking only to God, since people won't be able to understand you. You will be speaking by the power of the Spirit, but it will all be mysterious. 3 But one who prophesies strengthens others, encourages them, and comforts them.4 A person who speaks in tongues is strengthened personally, but one who speaks a word of prophecy strengthens the entire church.5 I wish you could all speak in tongues, but even more, I wish you could all prophesy. For prophecy is greater than speaking in tongues, unless someone interprets what you are saying so that the whole church will be strengthened.** So sometimes there is a strange language that sometimes comes from prayer or fasting. Not every church believes in that. There are certainly churches that embrace this. Remember

James 2:14 say **17 So you see, faith by itself isn't enough. Unless it produces good deeds, it is dead and useless. So you see if you're a church that doesn't believe in speaking in tongues**

m. **Gift of helping,** just like what you say, gift of helping people. The Bible specifies it towards the old and poor. Since becoming a Christian, have you had a bigger heart for the poor and elderly. With your church, are you always signing up for events for the elderly or the poor? This maybe a sign that you are maturing as a Christian.

Do you love God? Do you love your fellow man? A sign of being a Christian is loving God. As you start to read the Bible you find out how much God loves you. As you pray to God you see how much God loves you.

*Luke 10 25-27***25 One day an expert in religious law stood up to test Jesus by asking him this question: "Teacher, what should I do to inherit eternal life?"26 Jesus replied, "What does the law of Moses say? How do you read it?"27 The man answered, "'You must love the Lord your God with all your heart, all your soul, all your strength, and all your mind.' And, 'Love your neighbor as yourself.'"** Jesus thought this was the most important commandment, more than anything, second being to love your fellow man.

As you follow Christ and read the word, you realize how much sin you have created against other people. So, since you have sinned and realize your faults. You are more likely to forgive someone else's sins. When you come to Christ. For a lot of Christians, the hate inside and bad emotions go away. So there is room for love. As Jesus stated in Luke, the second biggest commandment was to love thy neighbor.

Conclusion

How can you tell if you're a good Christian? There are many ways to tell. Are you hearing from God? Do you have a good understanding of the Bible? Have you seen any miracles? Do you see things from God'S perspective? Do you have gifts of the spirits (Prophecy, Evangelist,

Pastoring, teaching, spiritual gifts, giving, words of wisdom, words of knowledge, healing, discerning spirits, gift of tongues, and the gift of helping)

15. WHAT CAN YOU DO TO BECOME A GOOD CHRISTIAN?

We talked about what it looks like when you are a good Christian. Here are 12 things to increase your faith.

1. The very first step is to acknowledge that you have sinned. *Romans 2 23-24* **23 For all have sinned, and come short of the glory of God;24 Being justified freely by his grace through the redemption that is in Christ Jesus. We have to come to done wrong. In order to do right, we need to acknowledge our wrong. Repent from your sins.**

2. The second thing is to surrender to Christ. *Psalm 37:7* **Rest in the LORD and wait patiently for Him; Do not fret because of him who prospers in his way, because of the man who carries out wicked schemes. You surrender to God. Because to become a Christian, you can not do it by yourself. Being a Christian is not about following rules, but a new lifestyle.**

3. *Mark 12 :30* **And you must love the Lord your God with all your heart, all your soul, all your mind, and all your strength." So, despite in your mind believing GOD God with all your heart, mind, and sou.**

4. Read the Bible and follow what the Bible says 2nd *Timothy 3:16* **16 All scripture is given by inspiration of God, and is profitable for doctrine, for reproof, for correction, for instruction in righteousness:. Also read study guides and other biblical book which will pull out more concepts.**

5. Get trained in things of God. Got to Bible study go to church, go to christian events, It's not enough to read the Bible. It's good to hear what other people faith is like. You could see how God has answered prayers and you could see how seasoned people break down scripture.

6. Allow God to change you. As you read the Bible and ask God for transformation. Understand that as you do these things, things will change. The friends you had may say you've changed. Accept the change

7. Be accountable for your actions. Stay on the path; if you veer off, get back on. That's why it's important to have someone who could be a disciple to you. Someone that you could seek counsel from if you have a question about God or what to do in a situation

8. Understand that things may not go as you like. *2 Timothy 3:12* **12 Yes, and everyone who wants to live a godly life in Christ Jesus will suffer persecution. Jesus was without sin, yet he still did not have every prayer answered. He still suffered pain and suffering. He still had hopes and dreams that were unfulfilled. So, being a follower of Christ, you will have the same.**

9. *Proverbs 3 5-6* **Trust in the Lord with all your heart; do not depend on your own understanding.6 Seek his will in all you do, and he will show you which path to take.** So don't believe in what you could see but believe in God word. They are going to be people who question God. Or when you have an experience with God, they will downplay it.

10. Pray on a daily basis. Expectation is the breeding ground for miracles. *Mark 11 22-24* NLV **Then Jesus said to the disciples, "Have faith in God. 23 I tell you the truth, you can say to this mountain, 'May you be lifted up and thrown into the sea,' and it will happen. But you must really believe it will happen and have no doubt in your heart. 24 I tell you, you can pray for anything, and if you believe that you've received it, it will be yours**.

11. Everything should point to Christ. As you transform, your radio stations should point to Christ, the TV shows should point to Christ, and the friends you hang out with should point to Christ. It's important

12. You should spread the gospel *Mark 16 15-17* **And then he told them, "Go into all the world and preach the Good News to everyone. 16 Anyone who believes and is baptized will be saved. But anyone**

who refuses to believe will be condemned. 17 These miraculous signs will accompany those who believe: They will cast out demons in my name, and they will speak in new languages. You should not only spread the Bible, but also disciple a few people. Jesus spread the Bible through many lands, but the 12 disciples he personally interacted with. 12 people he spent the whole day and night with. Jesus grew these people spiritually. Without spending time with these we wouldn't have any of the New Testament. So it just like Jesus, we need to strengthen up.

Conclusion

How do you become a better Christian? Did you acknowledge your sins? Did you submit to God? Do whatever you want to do. Do your actions meet the Bible? Is your heart towards God? Are you reading the Bible? Are you finding counsel when the Bible is unclear. Are you getting trained by the things of God?

Are you allowing God to change your heart, are you held accountable? Jesus said if you are a follower of Christ, you could expect to be persecuted. Are you praying on a daily basis? Are you spreading the gospel of Jesus Christ? Are you trusting God with all your might?

Most importantly, do you understand that Christian is an action? Just because you understand that Jesus is real and he died for sins, that's just a start. Just like in grade school, as you grow in the Lord, there are grades. Unlike grade school, there is no level to finish.

16. WHY ARE MARRIAGES FAILING?

Marriages are failing at an increasing rate. The latest statistics show that 60% of marriages are set to fail. People are more scared of getting into marriage because they had parents who were divorced and doing that. For the first time, there are more single people than married people. So why are marriages failing, and what could we do about it?

The first reason has to do with pride. *Proverbs 16 5* **5 Pride disgusts the Lord. Take my word for it—proud men shall be punished. Pride says that I shouldn't have to do this. Pride was the downfall of the devil. The devil or Lucifer was the opening act when it came to God.** *Ezekiel 28 1-9.* **Then this message came to me from the Lord: 2 "Son of man, give the prince of Tyre (Satan) this message from the Sovereign Lord: "In your great pride you claim, 'I am a god! I sit on a divine throne in the heart of the sea.' But you are only a man and not a god, though you boast that you are a god.3 You regard yourself as wiser than Daniel and think no secret is hidden from you.4 With your wisdom and understanding, you have amassed great wealth—gold and silver for your treasuries.5 Yes, your wisdom has made you very rich, and your riches have made you very proud.6 "Therefore, this is what the Sovereign Lord says: Because you think you are as wise as a god,7 I will now bring against you a foreign army, the terror of the nations. They will draw their swords against your marvelous wisdom and defile your splendor!8 They will bring you down to the pit, and you will die in the heart of the sea, pierced with many wounds.9 Will you then boast, 'I am a god!' to those who kill you? To them, you will be no god but merely a man.**

Let's break this down. So God is talking about Lucifer. He is saying that Lucifer boasts that he is a God but he is beneath God. That Lucifer claims to be smarter to Daniel. Again God says satan is not God. It is pride that thinks that he smarter than God or even on God level.

In *Ezekiel 28 1*1- **Then this further message came to me from the Lord: 12 "Son of man, sing this funeral song for the king of Tyre. Give him this message from the Sovereign Lord."You were the model of perfection, full of wisdom and exquisite in beauty.13 You were in Eden, the garden of God. Your clothing was adorned with every precious stone[b]— red carnelian, pale-green peridot, white moonstone blue-green beryl, onyx, green jasper, blue lapis lazuli, turquoise, and emerald—all beautifully crafted for you and set in the finest gold. They were given to you on the day you were created.14 I ordained and anointed you as the mighty angelic guardian.[You had access to the holy mountain of God and walked among the stones of fire. You were blameless in all you did from the day you were created until the day evil was found in you.16 You're rich commerce led you to violence, and you sinned. So I banished you in disgrace from the mountain of God. I expelled you, O mighty guardian, from your place among the stones of fire.17 Your heart was filled with pride because of all your beauty. Your wisdom was corrupted by your love of splendor. So I threw you to the ground and exposed you to the curious gaze of kings.**

So from the scripture, you could see that Satan was made with Jewels and precious stones. That he was the guardian of the angels. He had access to God mountain. He was filled with pride because of your beauty. Wisdom was corrupted.

Pride was the original sin. When Adam and Eve were in the Garden, they were told not to eat from the Tree of the Garden (Genesis 2:15-17). **The Lord God placed the man in the Garden of Eden to tend and watch over it. 16 But the Lord God warned him, "You may freely eat the fruit of every tree in the garden— 17 except the tree of the knowledge of good and evil. If you eat its fruit, you are sure to die."** Then in *Genesis 21-22* God made women **21 So the Lord God caused the man to fall into a deep sleep. While the man slept, the Lord God took out one of the man's ribs[d] and closed up the opening. 22 Then the**

Lord God made a woman from the rib, and he brought her to the man. So God told man not to eat the forbidden fruit, then God made women. So God never tells women about the forbidden fruit.

In *Genesis 2 3-6 s*ay the following: **The serpent was the shrewdest of all the wild animals the Lord God had made. One day he asked the woman, "Did God really say you must not eat the fruit from any of the trees in the garden?"2 "Of course, we may eat fruit from the trees in the garden," the woman replied. 3 "It's only the fruit from the tree in the middle of the garden that we are not allowed to eat. God said, 'You must not eat it or even touch it; if you do, you will die.'"**

4 "You won't die!" the serpent replied to the woman. 5 "God knows that your eyes will be opened as soon as you eat it, and you will be like God, knowing both good and evil."6 The woman was convinced. She saw that the tree was beautiful and its fruit looked delicious, and she wanted the wisdom it would give her. So she took some of the fruit and ate it. Then she gave some to her husband, who was with her, and he ate it, too.

I think there are so many people who miss this. Adam was with Eve when the serpent was convincing Eve to eat the forbidden fruit. Adam was told directly by God not to eat this fruit. Eve at most hears this from second knowledge about the fruit of knowledge according to the Bible.

In *Genesis 3:11-13* he replied, **"I heard you walking in the garden, so I hid. I was afraid because I was naked."[11] "Who told you that you were naked?" the Lord God asked. "Have you eaten from the tree whose fruit I commanded you not to eat?"[12] The man replied, "It was the woman you gave me who gave me the fruit, and I ate it."[13] Then the Lord God asked the woman, "What have you done?""The serpent deceived me," she replied. "That's why I ate it."**

14 Then the Lord God said to the serpent, "Because you have done this, you are cursed more than all animals, domestic and wild. You will crawl on your belly, groveling in the dust as long as you live.[15]

And I will cause hostility between you and the woman, and between your offspring and her offspring. He will strike[b] your head, and you will strike his heel."[16] Then he said to the woman, "I will sharpen the pain of your pregnancy, and in pain you will give birth, and you will desire to control your husband, but he will rule over you.[c]"[17] And to the man he said, "Since you listened to your wife and ate from the tree whose fruit I commanded you not to eat, the ground is cursed because of you. All your life, you will struggle to scratch a living from it.[18] It will grow thorns and thistles for you, though you will eat of its grains.[19] By the sweat of your brow will you have food to eat until you return to the ground from which you were made. For you were made from dust, and to dust you will return."

Pride tells us that you are better than everybody. Pride says that you don't have to change. Pride says that you are better than that person. You get into an argument, and you don't feel like you need to change. That was the downfall of Satan. He felt he was better than God. He felt like he didn't have to listen. He felt like he was above God. So did Adam feel like he didn't have to listen to God. When confronted by God instead of taking responsibility, he blamed Eve. Adam was there when God told him it would be dead if he ate from the Tree of the knowledge. Yet he chose to disobey God because he felt he knew better than God. He was not accountable for his actions either.

So in marriages, we need to do the opposite. We need to be humble and unprideful. What that looks like instead of blame-finding or not taking ownership. Everyday you wake up and figure out what things you can do bless your spouse. It's about taking responsibility for your role. It's not 50/50 but 100%. Regardless of what your spouse does, you do everything you can to make things right. We need to be tender to their heart.*1 Peter 5:6* **6 Therefore humble yourselves under the mighty hand of God [set aside self-righteous pride], so that He may exalt you [to a place of honor in His service] at the appropriate time**. God wants us to be humble. Marriage is a compromise, instead of looking to get what you want. Look for what you could do to get as many things as your spouse would want. Look to see as many things you could do that require much effort. By doing this, it makes your spouse do the things you want.

The second thing they marriages are lacking is acceptance. We all have fallen short of what we *Matthew 6 14-15 NLV* [14] **For if you forgive other people when they sin against you, your heavenly Father will also forgive you. [15] But if you do not forgive others their sins, your Father will not forgive your sins. We need to forgive whatever your spouse has done. It takes more energy to hold onto the anger than to forgive**. Love shouldn't be dependent on what someone does or doesn't do. Think of your spouse as kid or child. Your kid may do things you don't approve of, or your kids may not do what you expect of them. They may do what you expect of them. Your kids love is not conditional. You don't say, Billy. Since you don't clean your room, you can eat dinner. Or, Billy, you left the milk out, so you can't stay here tonight. No matter what your child does, you find a way to forgive them. You find a way to accept what they've done. You fight through of what they don't do. You don't forget what they did, but you find a way to accept it and forgive it.

Matthew 18 21-22 **Then Peter came to Jesus and asked, "Lord, how many times shall I forgive my brother or sister who sins against me? Up to seven times?"[22] Jesus answered, "I tell you, not seven times, but seventy-seven times. It doesn't matter how many times they done things, but we need to forgive.**

Also, understand that the person you marry is going to change. Think back 7 years,: were you at the same job? Did you have the same interests? Did you have any hobbies? Your body would have generated Red blood cells live for about four months, while white blood cells live on average more than a year. Skin cells live about two or three weeks. Colon cells die off after about four days. Sperm cells have a lifespan of only about three days, while brain cells typically last an entire lifetime. We are literally totally different beings in 7 years, so in order to have a long-lasting change. We need acceptance.

The third reason why marriages are failing is words. Words are important. Words should uplift rather than put down. *Proverbs 18:21* **Death and life are in the power of the tongue, and those who love it will eat its fruits.**

To demonstrate how important words are, I wanted to share with an experiment. A guy had three plants. One plant said hateful words. Hateful words like I hate you, you're ugly you're a terrible plant and other words. One plant he said nothing to. The last plant he said positive things. Positive words like "your beautiful," "you're a great plant," and others. The plant that he said hateful words to did not grow; it withered and died. The second plant that he stated nothing there was some growth. The last one he said there was much growth, and the one they said nothing. There are more experiments out there. There are experiments with rice cereal and other experiments. All lead that when words are negative, there is death or slow growth. When nothing is said, there is slow growth or no growth. When talking positive words that showed the most growth.

Remember what the Bible says about marriage. *Matthew 19 4-6* **Jesus replied. "They record that from the beginning, 'God made them male and female.'"** [5]**, And he said, "'This explains why a man leaves his father and mother and is joined to his wife, and the two are united into one.'** [6] **Since they are no longer two but one, let no one split apart what God has joined together. So as Jesus said when you they became husband and wife they become one flesh. So when you use hurtful words you hurt yourself.**

If your husband is always on the couch watching TV. You may think that "*you're so lazy just watching TV, why don't you take out the trash or do something.*" Instead of that say "you work so hard, I could understand why you need your rest. When you get a chance, I would appreciate it if you could take out the trash. Or if you have a wife who is always late. You may think that "*Can't we get there on time, don't take too long. I am leave you if you take too long.*" Instead, wow it was well worth the wait when I see you get ready. Next time, can we get ready a little bit early? What can I do help you get ready.

The next reason marriages are failing is a lack of boundaries or no boundaries. In your home is the inner circle. Everyone outside your home is outside the circle. You must be careful who lets you stay at their home. They may seem like they want the best for you. There are people who actually come to kill, destroy, and steal your relationship. *Genesis 2:24* says **Therefore a man shall leave his father and his mother and hold fast to his wife, and they shall become one flesh.**

As married couples, we must think of ourselves as a unit. If there is a friend of the husbands, then that friend should also be a friend of the husbands, and that friend should also be a friend of the wives. You guys are a unit any friends of the family should know both the husband and wife. If your wife doesn't like to be spend too much time with your in-laws. Then you negotiate that with the understanding your spouse comes first. There is a hierarchy of things with God being first and spouse coming second. Everyone else should fall behind that line.

If there is a conflict between the husband, the friend, and the wife. Then it is the husband job to make things right. With the understanding that, if a choice had to be made between his wife and his friend, his wife would be first. Not to say that the wife can't be wrong but your loyalty to your wife can't be questioned. You have to live in harmony with your wife. You don't live with your friend.

If there is conflict between husband and wife family. Then your wife job is to make things right with the family. With the understanding that the husband should be put a head above the family. Now you should not disown your family, but you might have limit their time. The wife should manage that situation.

The fifth reason why marriages are failing. The reason is that people are not fighting fair. What does that mean? How do you fight fair?

- The first thing to do to fight fair is to stay on topic. Do you ever have a fight and bring up past fights? Do you ever get into a fight, and when your spouse accuses you of something, instead of accepting blame do you blame them for the same thing? When you are being verbally attacked, do you listen to what is being said or to how it is being said? Do you ever say you're being loud or don't yell at me.
- The second thing that is important when fighting is how you feel after the fight. Are voices raised? It's important that people are heard, it's important that voices are not loud. It's important that there be no name-calling. Writer Gerald Jampolsky had a saying, "You could be right, or you could be happy." When you argue and say you're right, that means you are wrong.
- Taking this a step further about being right and wrong. God has made you beautifully made. Everybody is so unique, nobody is the same. If you have a wife who is always early and hates being late. You have a husband who always goes late. This husband and wife always argue about being late or early. Instead of the wife and husband focusing on whether you are right or wrong. They should focus on each being different. Instead of the husband saying the wife is wrong for wanting to be early. Instead of the wife saying the husband is wrong for not wanting to be too early. Our differences don't make marriages worse, they make us better. That is why God made us different. You don't want to marry yourself, that's why you marry someone different. Ephesians 4 2-3 NLV says **Always be humble and gentle. Be patient with each other, making allowance for each other's faults because of your love. 3 Make every effort to keep yourselves united in the Spirit, binding yourselves together with peace.**
- When we fight, you gotta understand something, you are not going to change your husband or wife. The only person you can change is yourself. You could pray for God to change your partner's heart.

Remember, marriage is a series of compromises. So these are important things when arguing.

So the 6th and most important reason why marriages are failing is lack of God. God never meant for us to go through life without him. Psalm 32:8 The LORD says, **"I will guide you along the best pathway for your life. I will advise you and watch over you." There is nobody smarter than God**. Ezekiel 36 :26 says **Then it will be as though I had sprinkled clean water on you, for you will be clean—your filthiness will be washed away, your idol worship gone. 26 And I will give you a new heart—I will give you new and right desires—and put a new spirit within you. I will take out your stony hearts of sin and give you new hearts of love. 27 And I will put my Spirit within you so that you will obey my laws and do whatever I command**.

As you read his word, pray and do things of God. You will start to transform. As you draw closer to God, you start to transform from an old self into a new, improved self. It would be less about you and more about your mate. You will be slower to anger, and you will be more tender-hearted towards each other and quick to forgive. *Colossians 3 12 -13* say. **Since you have been chosen by God, who has given you this new kind of life, and because of his deep love and concern for you, you should practice tenderhearted mercy and kindness to others. Don't worry about making a good impression on them, but be ready to suffer quietly and patiently. [13] Be gentle and ready to forgive; never hold grudges. Remember, the Lord forgave you, so you must forgive others.**

Remember God could see the future in *Jeremiah a 29* [11] **For I know the plans I have for you, says the Lord. They are plans for good and not for evil, to give you a future and a hope. [12] In those days, when you pray, I will listen· [13] You will find me when you seek me, if you look for me in earnest.** So since God could see what our future is. He could see what would be the best path for us. So seek God whenever you are making a big decision. Also God will set provisions ahead of time because he knows what you need

Conclusion

We tried to tackle a big problem in the world. The destruction of marriages, why are so many marriages failing? One of the reasons we looked at was Pride. That you don't need any help. You can get repairs on your home and your car, but when it comes to marriage, you don't want any help. You could handle things on your own. You have to ownership don't worry about whether the spouse is doing their part. You need to be the best husband or wife.

When fighting with a spouse, we need to stay on topic; it's important to hear the other side and embrace your differences. You don't want to marry yourself, your partner has different experiences. That's what makes marriage so great. The last thing you also want to focus on when fighting is that your spouse is not going to change. You have to be the one to change.

Also what marriages are missing now is forgiveness. All of us have fallen short. We need to have a forgiveness in our hearts. We have all have done something wrong to another individual. We have been done to by something we didn't like. Marriages is going to be full of forgiveness. You forgive that person then it should not be brought up again.

In marriage, as in life, your words are going to speak life or death. You should never speak bad about your husband, even if he is that person. If he doesn't take out the garbage, don't call him lazy, That is not going to motivate him to do that. You could ask him to take out the garbage. Name calling has no place in a good marriage. You want to say Good things about the spouse. You want to inspire the wife or husband to do better.

As a marriage, you need God. God came up with marriage since the days of Adam and Eve. God wants your marriage to succeed. When you have in your life, you can access untapped knowledge. Who better knows your spouse than the creator of your husband and wife? You need God to overcome obstacles that may come. There are no obstacles that God can't overcome.

17. IS THERE A HELL, AND WHY DOES GOD ALLOW HELL?

Why would a loving God wants us to perish? John 3:16 says **For God loved the world so much that he gave his only Son so that anyone who believes in him shall not perish but have eternal life. [17] God did not send his Son into the world to condemn it, but to save it.**

God does not get any pleasure from sending someone to hell. But because God is so good, he can't be around evil. Think of a cup of water, and think of sin as oil. They are both liquids, but there is clearly a difference between the two. They naturally separate. In *Habakkuk 1:13*, b**ut you are pure and cannot stand the sight of evil. Will you wink at their treachery? Should you be silent while the wicked swallow up people more righteous than they?**

At one time, Satan was one of God's most trusted people. In *Ezekiel 28:2,*

So Satin was a trusted member. He was one of the most important angels. Jesus talks about Satan falling from heaven. In *Luke 10:18-19,* **18 "Yes," he told them, "I saw Satan fall from heaven like lightning! 19 Look, I have given you authority over all the power of the enemy, and you can walk among snakes and scorpions and crush them. Nothing will injure you**.

One thing about God is that he is holy. Holy translates to "cut off" or seperate. When Satan started to turn bad. God can not be around that anymore. See, God is pure and he is good. He does not wink at evil.

That is why there is a hell. There is a separation from good and evil. Think of God as water and evil as oil. *Romans 6 : 22-23 NLV* **22 But now you are free from the power of sin and have become slaves of God. Now you do those things that lead to holiness and result in eternal life. 23 For the wages of sin is death, but the free gift of God is**

eternal life through Christ Jesus our Lord. We have all done things that have deserved separation from God. The devil was cast out because he thought he was better than God. As he got separated he got even worse and worse.

If you look at the Bible hell and the devil wasn't really mentioned in the Old Testament. Currently, right now there are many Jewish people who don't believe in hell. From the day that Adam disobeyed God there has been a separation. The separation has gotten bigger and bigger. *John 3:16* "**For this is how God loved the world: He gave his one and only Son, so that everyone who believes in him will not perish but have eternal life. 17 God sent his Son into the world not to judge the world, but to save the world through him.**

When there is separation from God we no longer have God graces anymore. In *Romans 8 :31* **31 What shall we say about such wonderful things as these? If God is for us, who can ever be against us**? When we sin, we turn our back against God. We go against the most powerful thing in the whole world. When is God of us? In *Numbers 22* 5-**So Balak, king of Moab, 5 sent messengers to call Balaam son of Beor, who was living in his native land of Pethor near the Euphrates River. His message said:**

"Look, a vast horde of people has arrived from Egypt. They cover the face of the earth and are threatening me. 6 Please come and curse these people for me because they are too powerful for me. Then perhaps I will be able to conquer them and drive them from the land. I know that blessings fall on any people you bless, and curses fall on people you curse."

7 Balak's messengers, who were elders of Moab and Midian, set out with money to pay Balaam to place a curse upon Israel. They went to Balaam and delivered Balak's message to him. 8 "Stay here overnight," Balaam said. "In the morning, I will tell you whatever the Lord directs me to say." So the officials from Moab stayed there with Balaam. 9 That night God came to Balaam and asked him,

"Who are these men visiting you?10 Balaam said to God, "Balak son of Zippor, king of Moab, has sent me this message: 11 'Look, a vast horde of people has arrived from Egypt, and they cover the face of the earth. Come and curse these people for me. Then perhaps I will be able to stand up to them and drive them from the land.'12 But God told Balaam, "Do not go with them. You are not to curse these people, for they have been blessed! So Balak, king of Moab, 5 sent messengers to call Balaam son of Beor, who was living in his native land of Pethor[a] near the Euphrates River. His message said:

"Look, a vast horde of people has arrived from Egypt. They cover the face of the earth and are threatening me. 6 Please come and curse these people for me because they are too powerful for me. Then perhaps I will be able to conquer them and drive them from the land. I know that blessings fall on any people you bless, and curses fall on people you curse."

7 Balak's messengers, who were elders of Moab and Midian, set out with money to pay Balaam to place a curse upon Israel. They went to Balaam and delivered Balak's message to him. 8 "Stay here overnight," Balaam said. "In the morning, I will tell you whatever the Lord directs me to say." So the officials from Moab stayed there with Balaam.9 That night God came to Balaam and asked him, "Who are these men visiting you?"10 Balaam said to God, "Balak son of Zippor, king of Moab, has sent me this message: 11 'Look, a vast horde of people has arrived from Egypt, and they cover the face of the earth. Come and curse these people for me. Then perhaps I will be able to stand up to them and drive them from the land.'"12 But God told Balaam, "Do not go with them. You are not to curse these people, for they have blessings.

So even if people may curse you, they can't harm you. But if your turned your back on God. They can and do harm to you. That's why there is a hell because these are people who have turned their back on God. That is why there is a hell.

God remedy for the separation was to have Jesus Christ. No matter how much sin you have, if you have Jesus Christ in your heart, you can have everlasting life with GOD. *John 14 6-7* **Jesus told him, "I am the way, the truth, and the life. No one can come to the Father except through me. 7 If you had really known me, you would know who my Father is. From now on, you do know him and have seen him!"** By inviting Jesus in your heart it breaks down that separation that started with Adam. Some people have described Jesus as the second Adam.

As the separation grew from growing sin, Hell grew bigger and bigger. That is why there is so much more information about hell and satan. Because he became from a minor character to a major character in the Bible. He continues to grow from when Jesus Christ came on this earth over 2000 years ago today. As it describes in the Bible there will be a judgement day where everyone will be judged.

Revelation 20 11-15 is **11 Then I saw a great white throne and him who was seated on it. The earth and the heavens fled from his presence, and there was no place for them. 12 And I saw the dead, great and small, standing before the throne, and books were opened. Another book was opened, which is the book of life. The dead were judged according to what they had done as recorded in the books. [13] The sea gave up the dead that were in it, and death and Hades gave up the dead that were in them, and each person was judged according to what they had done. [14] Then death and Hades were thrown into the lake of fire. The lake of fire is the second death. [15] Anyone whose name was not found written in the book of life was thrown into the lake of fire.**

So even know there is hell. God never wanted this separation from God. God last remedy to combat this. Is to end the world as we know it. There won't be anybody to deny God's existence. There will be people who accept God's rule, and there will be people that will be burned in the eternal fire.

Conclusion

God does not seek any pleasure in seeing anyone go to hell. At the same time his only son went to hell for our sins. If he allows his son to go to hell than died a sinners death. Even know he was without any sin. Then why not us.

Satan was a trusted angel who boasted that he was God. Satan was cast out of heaven. Satan was separated from God. This separation between Satan and God got bigger. This is because Satan went from a trusted assistant to an enemy of God. As satan became more evil, he convinced man to sin or to separate from God. God's remedy to that is for his only son to die a sinner's death

In the Bible we can not pick and choose which Bible verses we follow. We must take the Good and the bad. It's maybe unpleasant to talk about hell. It's much more pleasant to talk about heaven. There has been a series of movies that talk about heaven. What we don't have are movies about visiting hell. There are people who have been reported as going to hell.

18. WHY DOES SATAN TRY TO KILL, STEAL, AND DESTROY? WHY DOES HE NOT LIKE HUMANS

Some people would say that satan was always evil. The scriptures say that Satan was a trusted top Angel. *Isaiah 14-15* **You were the anointed cherub who covers; I established you; You were on the holy mountain of God; You walked back and forth in the midst of fiery stones. [15] You were perfect in your ways from the day you were created.**

Now, clearly can't get into the mind of Satan or the devil. We could infer from scripture what he was thinking. Genesis 3 1-7 says **Now the serpent was more crafty (subtle, skilled in deceit) than any living creature of the field which the Lord God had made. And [the serpent (Satan) said to the woman, "Can it really be that God has said, 'You shall not eat from any tree of the garden'?" [2] And the woman said to the serpent, "We may eat fruit from the trees of the garden, [3] except the fruit from the tree which is in the middle of the garden. God said, 'You shall not eat from it nor touch it, otherwise you will die.'" 4 But the serpent said to the woman, "You certainly will not die! [5] For God knows that on the day you eat from it, your eyes will be opened [that is, you will have greater awareness], and you will be like God, knowing [the difference between] good and evil. [6] And when the woman saw that the tree was good for food, and that it was delightful to look at, and a tree to be desired in order to make one wise *and* insightful, she took some of its fruit and ate it; and she also gave some to her husband with her, and he ate. [7] Then the eyes of the two of them were opened [that is, their awareness increased], and they knew that they were naked; and they fastened fig leaves together and made themselves coverings.**

So Satan wanted to tempt man. He wanted us to eat from the tree of knowledge. He wanted man to separate from God. He wanted us to have knowledge of Good and Evil. Why did Satan want to do that to man? The scriptures may shed light on that *Hebrews 1 3-4* **The Son radiates God's own glory and expresses the very character of God, and he sustains everything by the mighty power of his command. When he had cleansed us from our sins, he sat down in the place of honor at the right hand of the majestic God in heaven. 4 This shows that the Son is far greater than the angels, just as the name God gave him is greater than their names.**

In *Genesis 1 :27* **Then God said, "Let us make human beings in our image, to be like us. They will reign over the fish in the sea, the birds in the sky, the livestock, all the wild animals on the earth, and the small animals that scurry along the ground."**

That is an important line we are made in the image of your father. So we are able to make things like God. We are able to make decisions, and we are able to create things.

John 3 16 -21 **"For this is how God loved the world: He gave his one and only Son, so that everyone who believes in him will not perish but have eternal life. [17] God sent his Son into the world not to judge the world, but to save the world through him.[18] "There is no judgment against anyone who believes in him. But anyone who does not believe in him has already been judged for not believing in God's one and only Son. [19] And the judgment is based on this fact: God's light came into the world, but people loved the darkness more than the light, for their actions were evil. [20] All who do evil hate the light and refuse to go near it for fear their sins will be exposed. [21] But those who do what is right come to the light so others can see that they are doing what God wants."**

So I think there was some jealousy it is clear that angels were created before man. Satan was in the garden and cast down from heaven. Satan saw that Adam was cast into God'S image. I could imagine that may make Satan even more upset, Here you are casted down from heaven, and this new person comes on the scene. They look like God the person who cast you out. Then you deceive man.

Fast forward to the times of Jesus. You've been doing pretty well at messing up the earth. You finally get your revenge. You help kill God's son, Jesus Christ. Only to find out when Jesus is killed. Man is given a gift. A movement is started to reconcile man to God. This was never offered to Satan. Satan was God closest people, and Satan is kicked out of heaven. I am sure this would make Satan even more upset. Man is given a gift that doesn't have to be earned. All we have to do is invite him in.

I would imagine that Satan would do everything in its power to try to stop that reconciling process, satan has a hate for us. Satan does not want to know God or love God.

"No one can serve two masters. For you will hate one and love the other; you will be devoted to one and despise the other. You cannot serve God and be enslaved to money. You can't serve two masters.

Satan knows that his time is up. He is aware of the scriptures. He is trying to kill as many people as he can and persuade as many people as he can that there isn't Goda God. These are the last days. As we described in another question, we don't have many days left.

Conclusion

Have ever wondered why Satan is so bad? How can one of the top Angels go down to be the worst bad guy? It started with Satan having pride. Believing that he was more than what he says. That banished Satan from heaven.

When man was created, we were like God. We were created in God image. So why not Satan try to get back at God. So satan tempted man, and we fell from Grace. Satan then tempted the Son of God Jesus. Jesus did not fall to satan. Satan helps kill Jesus. Only to find out that man is given a gift. Man is given of Salvation *Romans 10 9:10* **If you declare with your**

mouth, "Jesus is Lord," and believe in your heart that God raised him from the dead, you will be saved. 10 For it is with your heart that you believe and are justified, and it is with your mouth that you profess your faith and are saved.

So, no wonder why Satan hates man. He thought that man would be on his side. That they would be separated from God. Jesus saved us by dying for our sins. We have a gift that satan or angels could have. If we believe in Jesus, we could have eternal life. Satan has been even more angry because he knows his days are numbered. This would be the last time for him to kill, steal, and destroy

19. WHAT IS THE STATE OF THE WORLD? IS THE WORLD IS GETTING WORSE?

What if we world was a village? What would that look like? If the world was a village of a 1,000 people. ⅔ of the world would be non-Christian. 95 % would be able to speak a different language other than english. 17% can not read and write. 23% don't have a shelter to protect from rain. 10 people would be dying from starvation, 150 people would be malnourished. 130 people wouldn't have access to clean water.160 people wouldn't have access to a toilet. 220 wouldn't have electricity. It paints a different picture than living in the United States. Things look different than what you may think. We are going to look at scientific reasons why the world is coming to an end.

1. Slavery In the world, there are more slaves in today's world than ever before in history. The slaves are treated worse than ever before. About half of the slaves 20 million are forced labor slaves. About 15 million people are forced into marriages they never wanted. There are also prostitutes and slaves. Also, about 10 million kids are forced into slavery. Whereas in 1850, it cost about 40,000 for a slave. The worldwide average for a slave is 90 dollars. There are many more slaves at a lower price.

2. Mass shootings In the United States, there has been a triple amount of mass shootings since 2011. There have been more than 90 mass shootings in the last 50 years. In 2016, there were over 300 people who were either injured or killed in mass shootings. Over 50,000 of people were injured by guns or killed by guns.

3. Income inequality There is more income inequality. The top 20% of Americans own 85% of the wealth in the United States. 1% owns more than 20% of the wealth. 60% of Americans own just 5% of the wealth. The income disparity got worse in 1974, the 1% owns about 9%. That income disparity has more than doubled in the last twenty or 30 years. Currently, the CEO earns 380 times the average worker's salary. What a CEO makes in an hour an average work month to get that same amount. It is said that income disparity is worse than it was in Medieval lives.

4.Running out of water. The world is running out of drinkable water. Water demand globally is projected to increase by 55% between 2000 and 2050. Right now, according to a NASA-led study, many of the world's freshwater sources are being drained faster than they are being replenished. Much of the same is happening in California. From 2011 to 2016, the state suffered its worst drought in 1,200 years. Its major aquifers receded at a combined rate of 16 million acre-feet per year, and roughly 1,900 wells ran dry. The researchers studied changes in water levels in 37 major aquifers worldwide between 2003 and 2013. Of those, 21 aquifers have declined, many of them in arid or semi-arid regions. Thirteen of those declining aquifers — about one-third of the total — were classified as "highly stressed," "extremely stressed," or "overstressed," with the most severe situations occurring in dry areas where little or no water seeps into the ground to offset the amounts pumped out. Some of the regions with the most pronounced groundwater depletion are also major farming areas. The aquifer beneath California's Central Valley, for instance, was labeled as highly stressed. The researchers found the Arabian Aquifer System, which supplies water to more than 60 million people, to be the most overstressed in the world, followed by the Indus Basin aquifer of northwestern India and Pakistan, and the Murzuk-Djado Basin in northern Africa."The red flags are that over half of the world's biggest aquifers are being depleted. They are past sustainability tipping points, and a third of those big aquifers — 13 of those — are seriously distressed," said Jay Famiglietti, a professor of earth system science at UC Irvine who co-authored the studies.

5. **Oil and Gas.** We use oil to run our cars and trucks. There are a lot more uses for that than just that. There are over 6000 products that involve oil. Some of the things would be Shower Curtains Guitar Strings Luggage Aspirin Safety Glasses Antifreeze Football Helmets Awnings Eyeglasses Clothes Toothbrushes Ice Chests Footballs Combs CD's Paint Brushes Detergents Vaporizers Balloons Sunglasses Tents Heart Valves Crayons

Parachutes Telephones Enamel Pillows Dishes Cameras Anesthetics Artificial Turf Artificial limbs Bandages Dentures Model Cars Folding Doors Hair Curlers Cold cream Movie film Soft Contact lenses Drinking Cups Fan Belts Car Enamel Shaving Cream Ammonia Refrigerators Golf Balls Toothpaste Gasoline Ink Dishwashing liquids Paint brushes and thousands of many more things. They say with our current use of oil we have about 50 years left of oil.

Natural Gas we use for heating, electrical plants, cooking, and other uses. They say there are about 58 years left of natural gas.

6. Phosphorous. You may not realize it, but many of our plants, fruits, and vegetables need fertilizer. One of the main ingredients in fertilizer is phosphorus. With the population ever increasing, food will always be needed. They estimate that with the current, it's less than 100 years.

7. Gold and silver. They estimated that we would have mined all the gold by 2030 and all the silver by 2035. Gold is used for Jewelry, electronics, and dental work. What makes Gold and silver worse is the commodities. Commodities is like the stock market, where you have a piece of paper that represents Gold and Silver. Right now, there are more people holding a piece of paper that represents Gold than all the Gold in the world. They estimate that range from say 100 commodities (paper) is equivalent to 1 ounce of real gold. Another estimate has 293 ounces of commodities (paper) is about the value of 1 ounce of real Gold. The current price of silver is about 20.00 a ounce (paper) but a real estimate places the true value is about 1200 an ounce. So the price of Gold and Silver is artificially low.

As a world, we are more dependent on each other in terms of the economy. So when Asian markets are in recession, that would affect American markets as well. So with that, if there was a recession and people were cashing in their commodities. That would force the owners of the commodities to pay out millions of dollars. It's got even worse where world commodity exports increased fivefold from 1995 to 2011 (when they reached US$6.055 trillion) and in 2011, the alternative trading system (ATS) of electronic trading featured computers buying and selling without a human dealer i. High-frequency trading (HFT) algorithmic trading is becoming more common. This setting is up for failure. With more money into commodities, it's going to get worse financially when things go bad. And with commodities being sold through computers, there is no human at the switch. So things could get worse.

8 **Natural disasters.** In 1970, the average number of natural disasters reported was 78; in 2005, this number jumped to 439. In 2010, natural disasters have affected more than 340 million people that year. 340 million is about the population of the United States. From 1980 to 2009, the number of climate-related disasters increased by 80 percent. Between 2001 and 2010, more than $1.2 trillion was lost to the increased rates of natural disasters. The intensity has increased in 2017; we had two 5-category storms and two category 4 storms. These are the most powerful storms. With the temperature rising, warm water is creating more storms, and those storms are getting more powerful.

The water is also melting at the polar caps. So, according to CNN, we lost 5 Samoan islands: Kale Island, Rapita Island, Zollies, Rehana Island, Kakatina Island. Whole islands are disappearing. According to CNN, we lost 5 Samoan islands: Kale Island, Rapita Island, Zollies, Rehana Island, and Kakatina Island. With more people in the world living near cities rather than the country.

9. Animals are disappearing. We are not sure about how many there are. Estimates vary on how many species there are. We have a better understanding of the stars than our own species. Estimates range from 2 million to 100 million species. The estimates on the species lost are between 2,000 and 100,000 species a day. The thing is, one of the species is gone, which has a chain effect on other species. Species are part of the ecosystem; when one dies off, it affects all animals.

For example, there has been a 50% reduction in bee colonies in the last 100 years. From 2007 to 2013, 10 million colonies were lost. Bees are responsible for pollinating approximately one-third of the United States' crop species, including almonds, peaches, apples, pears, cherries, raspberries, blackberries, cranberries, watermelons, cantaloupes, cucumbers, and strawberries. What worries a lot of people is the fact that pollination requires plants and we are not sure why bees are dying. Something has to do with fungicides, pesticides, and climate change.

There is a German amateur group called the Krefeld Entomological Society that monitors insect populations at 100 nature reserves in Western Europe and has been doing so since the 1980's. They discovered that by 2013, the population of insects in these reserves began to drop by an unbelievable near 80%. Without insects, there can't be animals that eat insects. When the insects aren't there, that means the frogs can't eat, and the birds can't eat. If frogs aren't there, snakes can't eat, and raccoons. If birds aren't there, wild dogs, coyotes, and bobcats can't eat.

9. Wars - There are estimates of about 190 countries in the world. There are about 130 armed conflicts currently. Most of the world is at war. In the world we live in, it's easier to talk about the countries that are not at war than the ones that are. These countries are known for staying neutral. Iceland, Switzerland, Costa Rica, Vatican City, New Zealand, and Ireland

The United States is one of the largest makers of War. The countries currently at war with Iran, Yemen, Somalia, Syria, and Yemen. Although we are not currently in Iraq and Afghanistan, there was a lot of deaths and money spent. These are the money, the impact, and the death of the United states recent wars.

A. Iran war / Conflict. It started in March 2026. With 900 air strikes in 12 hours. Killing over 1,100 Iraqi people. Among the people killed was Supreme Leader Ayatollah Ali Khamenei. These attacks also destroyed an all-girls school where 170 girls were killed. The wars so far have killed 13 US service men and are estimated to have displaced 3.2 million people. As of 3/21 the war has cost 18 billion dollars. With the Pentagon asking for an additional $200 billion.

C. Iraq: We have been involved for over 25 years 1.7 trillion, and over 4,000 us soldiers killed. With half a million civilians dead because of the conflict in Syria, we have been at war since 2009, with 4 Americans dead, and they spent 12 billion dollars. The United Nations estimated 400,000 Syrians have been killed because of the conflict.

D. Afghanistan: $783 billion is the amount we have spent. The United States Department of Defense lists 2,254 military personnel having died in Afghanistan, Pakistan, and Uzbekistan. Of these, 1,856 are due to hostile action and 398 are non-hostile. 31,000 civilians have died. 111,000 if you include civilians and military

E. Yemen: In the last couple of years, the United Nations has recorded a total of 13,000 civilian casualties, including 4,971 killed and 8,533 injured. Trump has made a recent arms deal worth $110 billion dollars. This is a change in the U.S. position on the Saudi-led coalition in Yemen's civil war. Many people argue that this is going to vastly increase civilian casualties.

F. Somalia: haven't been any recent American casualties. In 1992-1993, there were 29 people who died in the Somalia conflict, according to the Pentagon. More recently, when we engaged again with Solomian, there haven't been any deaths by our army. There have been two reporters who have died. In recent years, there have been over 50,000 casualties from this conflict. There are millions of people who need food or some sort of assistance. We have spent over $ 55 billion since 1992.

What would be surprising to most people is that the US has troops in 46 in 50 countries in Africa. In some capacities, they are sometimes training other troops or involved in combat. They estimate there has been 5,000 and 6,000 troops.

Since 2011, there have been 71 people who have died from terrorist attacks on American soil. Here are the numbers for how much we are spending on the military.

These are the top countries that spend on the military

1. The United States spend over 600 billion dollars a year on the military
2. China spends over 170 billion dollars on the military
3. Russia spends over 80 billion dollars on the military
4. Saudi Arabia spends over 60 billion dollars on the military
5. France spends over 60 billion dollars on the military.

I wanted to show these numbers to show how much the United States spends in military comparing to other countries. If you add the other 4 highest military-spending countries, you get a little bit more than half of what the United States spends on the military.

Compare that to how much we spend on the homeless in the United States: 45 billion. Even with spending that much money for the homeless people in 2014, there was an estimated over 500,000 people homeless on any given day. We could do better.

In the United States, we spend $ 70 billion on SNAP (The Supplemental Nutrition Assistance Program). 1 in 6 people in America face hunger. 49 million Americans struggle to put food on the table, More than 1 in 5 children is at risk of hunger. Among African-Americans and Latinos, it's 1 in 3. Even though we have this program

We have discussed in an earlier segment that in the United States, we have more of income inequality than ever before. So there are more people who are poorer. More people who need services include food and housing. What kind of world do we live in where we spend more on people killing people than on food for the needy or homeless? We need the military, and we need to defeat our enemies. When we spend billions of dollars to prevent people from dying from Terrorist attacks from happening since 2011, there has been 71 people who have died from Terrorist attacks. Is that worth all the money when people are homeless and when people don't have food to eat. We could still pay twice as much as the nearest country. And we still save 300 billion dollars. With 300 billion dollars, we could invest in new homes for the homeless. It is estimated that it would cost 75 billion dollars to provide free education. So we would have the money for that if we cut spending on military.

Abortions, God does not like any sin. *Romans 6:23* **For the wages of sin is death; but the gift of God is eternal life through Jesus Christ our Lord.** God especially does not like it when we go against the Ten Commandments. Remember that the Ten Commandments were the law that God gave to the Israelites. *Exodus 19 5* NLV **says Now if you will obey me and keep my covenant, you will be my own special treasure from among all the peoples on earth; for all the earth belongs to me.** [6] **And you will be my kingdom of priests, my holy nation.' This is the message you must give to the people of Israel**." Covenant means a promise or pledge, so God says if you keep my promise, you will be rewarded

In *Exodus 20 1- 3* **Then God spoke all these words:**[2] **"I am the Lord your God, who has brought you out of the land of Egypt, out of the house of slavery.**[3] **"You shall have no other gods before Me.** It continues *Exodus 20-13 Amp* [13] **"You shall not commit murder (unjustified, deliberate homicide).** There is something to be said when God verbally tells us not to do something. And before he speaks to the Israelites, he said, "If you keep my promise, you will be rewarded. That is why I think God is really saddened by the number of abortions.

There has been over 45 million abortions in the US since 1970, when abortion was legal to put that in perspective. Since that time period for all the wars that we have been in there has been about 8,000 deaths. When taking all the abortions deaths over the years in the US, it was more major genocides by Mao Zedong, Stalin's genocide, and Hitler's genocide.

Millennials were born from 1980 to 2000; they represent 75 million people. During 1980 to 2000, there were roughly 25 to 30 million abortions performed. This means that for one millennium died from an abortion out of 4 millennials did. One could argue that this generation is the most heavily aborted in human history. God has a plan for each of these children. When we abort them, we are taking away what God has designed for these people. Some of the most brilliant doctors, teachers, lawyers, and politicians came from humble backgrounds. Just because you have a limited income or are a single parent does not mean the child will come out badly.

Abortions have disproportionately affected blacks, people. The CDC reports that during the 1970's, about 24% of all U.S. abortions were performed on black women. That percentage rose to about 30% in the 1980's, and 34% in the 1990's. In the 2000s, it increased to about 36%. That means that about 31% of all U.S. abortions since 1973 have been performed on African American women. Estimates say that between 1973 and 2018, there have been around 60 million abortions in the United States, and 18 million of them were performed on black babies. In 2017, the black population in the U.S. stood somewhere around 40 million, which means that abortion has reduced the size of the black community by more than 30%—and that doesn't include the children and grandchildren that would have been born to those aborted more than a generation ago. Right now, there are more black babies aborted than born.

CoronaVirus 2019. CoronaVirus 2019. The world is facing one of the most dangerous viruses. 777 million people have been infected, with over 7 million people dead. In the United States, over 102 million people are infected with coronavirus, with 1.2 million deaths. What makes the virus so bad is that the majority of the 22 million people who have it cannot have any symptoms. Where the flu, you could tell you had the flu. You can't always tell that you have symptoms. The flu has a season from October to April. The coronavirus has no season,

To prevent further infections, we have shut down most economic activity. Worldwide, millions of people have lost their jobs. Millions of people have reduced pay or reduced

Hours. The major companies that have filed for bankruptcy is Hertz car rental, JCPenney, Pier

One, Models, True religion, J Crew, Neiman Marcus, GNC, Tuesday morning, Chuck E Cheese,

Brooks Brothers, New York company, Ann Taylor, California Pizza, Lord and Taylor, Men's Warehouse, Jos. A. Bank, K&G Fashions, and Stein Mart. You may have heard of some of the Companies.

In addition, according to the Chamber of Commerce. About 27 million small businesses

Have been affected by the coronavirus in some way. What makes this unique vs the flu is the lasting impact... We have a generation of kids. That is not in-person schooling. In-person schooling. We have a generation of people who do not know how to interact with peers. We have a generation of kids that may not know how to have Fun. We have a generation of people who are failing. This is not going away anytime soon. There are reports that, in Maryland, the number of F students has increased 6-fold due to the pandemic. If these students fail and drop out of school, it is said that this increases the chance of going to prison by 47%. We don't know what the long-term effects of this disease may be. Some people report "Brain fog." In addition, people may have issues with their heart. With the flu, there are no long term effects. There are some reports of hair loss and fatigue. These people may be affected by this and not know it. Some people also report tooth loss. There are reports of muscle pain and constant headaches long after thecoronavirus is gone. People who report insomnia can't sleep. This is going to affect us for a whole generation. Just as the Great Depression impacted a generation, so shall the Coronavirus. We won't know all the effects and the devastation until it's over. In the Bible they talked about at the end of days, there will be disease. Look at Luke 21 vs 9-119 When you hear of wars and uprisings, do not be frightened. These things must happen first, but the end will not come right away."10 Then he said to them: "Nation will rise against nation, and kingdom against kingdom. 11 There will be great earthquakes, famines and pestilences in various places, and fearful events and great signs

Conclusion

We live in a world where you could live in a bubble. Where you think everyone in the world looks like you, worships the same God you do. You may be surprised to learn that 95% can speak a language other than English. 20% of the world's population does not have access to electricity.

You may also be surprised at how bad this world is. Slavery is worse than it has ever been, where it has been. It's more widespread than it's ever been before. It's millions of people who have been enslaved.

It's also been a lot of mass shootings where over 50,000 people have been injured by a handgun. Income inequality has gotten worse. 1% owns 20% of the wealth. There hasn't been much income disparity since medieval times. Ceo now earns 380x the average worker's salary.

The world is running out of water. The earth is not replenishing the water. The demand for water increases, yet the supply decreases. It is said that supply is going to increase more than California, which is suffering the worst drought in 1,200 years. Oil, which runs more than cars and trucks, is stated to run out in less than 50 years. Oil makes Artificial Turf Artificial limbs Bandages Dentures Model Cars Folding Doors Hair Curlers Cold cream Movie film Soft Contact lenses Drinking Cups Fan Belts Car Enamel Shaving Cream Ammonia Refrigerators Golf Balls Toothpaste and those are just some of the things that oil make.

Phosphorus, which is the main source for fertilization. Phopperherous is said to run out in about 100 years. Gold and silver are running out. When they run out of gold and silver, this will disrupt many markets. Gold and silver are sold on exchanges more than what is out there. There will be a day of reckoning when gold is sold off, and its price will increase exponentially.

Natural disasters have increased by more than 80% in recent times. They are not only more of them, but also more powerful, causing death and destruction. We have lost whole islands due to natural disasters.

Animals are disappearing at an alarming rate. Estimates vary, but we lose anywhere between 2,000 and 10,000 species a day. When a species dies, so does that ecosystem. For example, bees are disappearing. Some estimates from 2007 to 2013 we lost 10 million colonies of bees. Bees are responsible for pollinating approximately one-third of the United States' crop species, including almonds, peaches, apples, pears, cherries, raspberries, blackberries, cranberries, watermelons, cantaloupes, cucumbers, and strawberries.

In the last question, we also looked at wars. Most of the world is at war. There are about 190 recognized countries. In the US alone, we are in 7 wars. Now these wars vary in involvement.

In Iraq, we have been involved for over 25 years 1.7 trillion, and over 4,000 US soldiers killed. Half a million civilians died because of the conflict in Syria. We have been at war since 2009, with 4 Americans dead. 12 billion dollars. The United Nations estimated 400,000 Syrians have been killed because of the conflict. The United States spends over 600 billion dollars a year for the military. China spends over 170 billion dollars on its military.

Compared to how much we spend on the homeless in the United States, it is 45 billion. Even with spending that much money for the homeless people in 2014, there was an estimated over 500,000 people homeless on any given day. We could do better. In the United States, we spend $ 70 billion on SNAP (The Supplemental Nutrition Assistance Program). 1 in 6 people in America face hunger. 49 million Americans struggle to put food on the table. More than 1 in 5 children is at risk of hunger. Among African-Americans and Latinos, it's 1 in 3. Even now, we have this program.

What kind of world do we live in where we spend more on killing people than on food for the needy or homeless? Is that worth all the money when people are homeless and when people don't have food to eat? We could still pay twice as much as the nearest country. And we still save 300 billion dollars. With 300 billion dollars, we could invest in new homes for the homeless. It is estimated that it would cost 75 billion dollars to have a free education. So we would have the money for that if we cut spending on the military.

What is more disturbing is the number of black people who have been aborted. It is estimated that because of abortions, the black population has decreased by 30%. The percentage has increased from the 1970's to 2000.

As you can see, the world is a messed-up place. It is even more reason for us to draw close to God. It's even more reason to get into the scriptures. It is not an exaggeration that the world is very troubled.

20. ARE WE IN THE END TIMES? IS THE WORLD ENDING?

Through the Bible there have always been prophets who could tell what was going to happen in the future. There were 12 major prophets and 12 minor prophets, all able to tell the future. Some of the things that were prophesied were coming from Jesus Christ we talked about earlier in the book. There have been hundreds of predictions of Jesus coming. We talked about some of them in the book, but some would say there are more than 300 instances in which Jesus was prophesied. There has been a prophecy about the end times. Some end times have gone true, and some have not yet come true. We will look at the prophecy regarding Israel, and we will look at other prophecies.

The Bible mentions Israel over 150 times. Mary Jesus' mom was mentioned in the Bible about 50 times. Israel was the promised land. This land was always sacred. This is the land where Jesus was born; this is where most of the prophets were born. *Genesis 26 2-5 Amp* **The Lord appeared to him and said, "Do not go down to Egypt; stay in the land of which I will tell you. [3] Live temporarily [as a resident] in this land and I will be with you and will bless *and* favor you, for I will give all these lands to you and to your descendants, and I will establish *and* carry out the oath which I swore to Abraham your father. [4] I will make your descendants multiply as the stars of the heavens, and will give to your descendants all these lands; and by your descendants shall all the nations of the earth be blessed, [5] because Abraham listened to *and* obeyed My voice and [consistently] kept My charge, My commandments, My statutes, and My laws."**

Isaiah was again one of the great prophets of the day. He was talking about Israel as a united country. Before 1948, Israel was just an idea. It had to be God hand for there to be a land for nothing. In *Isaiah 66:7-9, it talks* about Israel being created in one day. Zion was another word for Israel. **It**

reads the following: Before she (Zion) was in labor, she gave birth; Before her labor pain came, she gave birth to a boy." Who has heard of such a thing? Who has seen such things? Can a land be born in one day? Or can a nation be brought forth in a moment? As soon as Zion was in labor, she also brought forth her sons. "Shall I bring to the moment of birth and not give delivery?" says the Lord. "Or shall I who gives delivery shut *the womb*?" says your God.

Israel was created from a desert. It was a prophecy that God would transform the desert land into a "garden of Eden." *Isaiah 35:1-2* **The wilderness and the dry land will be glad; The Arabah (desert) will shout in exultation and blossom like the autumn crocus. It will blossom abundantly and rejoice with joy and singing. The glory of Lebanon will be given to it, the majesty of [Mount] Carmel and [the plain] of Sharon. They will see the glory of the Lord, the majesty *and* splendor of our God.** Israel is in the top ten exporters of flowers for the world.

In Isaiah's time, there was not a present Israel. He had a vision of Israel, which is now the modern Israel *Isaiah 60:4* **"Look and see, for everyone is coming home! Your sons are coming from distant lands; your little daughters will be carried home"** *Isaiah 60 8* **"And what do I see flying like clouds to Israel, like doves to their nests? They are ships from the ends of the earth, from lands that trust in me, led by the great ships of Tarshish. They are bringing the people of Israel home from far away."** This again speaks to the vast distances people would travel. It says from the ends of the earth. It even talks about flying in the sky (airplane travel). Tarshish was a far-off land.

In *Zechariah 12 2-9* **2 I will make Jerusalem like an intoxicating drink that makes the nearby nations stagger when they send their armies to besiege Jerusalem and Judah.3 On that day, I will make Jerusalem an immovable rock. All the nations will gather against it to try to move it, but they will only hurt themselves.[4] "On that day," says the Lord, "I will cause every horse to panic and every rider to**

lose his nerve. I will watch over the people of Judah, but I will blind all the horses of their enemies. 5 And the clans of Judah will say to themselves, 'The people of Jerusalem have found strength in the Lord of Heaven's Armies, their God.' 6 "On that day I will make the clans of Judah like a flame that sets a woodpile ablaze or like a burning torch among sheaves of grain. They will burn up all the neighboring nations right and left, while the people living in Jerusalem remain secure. 7 "The Lord will give victory to the rest of Judah first, before Jerusalem, so that the people of Jerusalem and the royal line of David will not have greater honor than the rest of Judah. 8 On that day, the Lord will defend the people of Jerusalem; the weakest among them will be as mighty as King David! And the royal descendants will be like God, like the angel of the Lord who goes before them! 9 For on that day I will begin to destroy all the nations that come against Jerusalem.

Some of this revelation has already happened, and some is yet to come. First, Israel is surrounded by Arab countries: Syria, Lebanon, Jordan, and Egypt. If we extend further, Iraq, Kuwait, Yemen, Morocco, Algeria. These are all in the Middle East. All are against Israel. There have been many conflicts between Israel and its neighbors, but I think nothing demonstrates God's unwavering love for Israel more than a 6 day war. Tensions were rising from the previous wars. Israel did a sneak attack and took out Egyptians air force. Nearly all the Egyptian planes were lost. They used ground troops and invaded the Gaza Strip and the Sinai. Syria and Jordan joined in attacking Israel.

In that short war, over 20,000 Arabs were killed, while only a hundred Israelis were killed. It is even more amazing when you look at all the people who were against Israel. You had Egypt, Syria, Jordan, Iraq, and Lebanon, which had troops and tanks against Israel. The Arabs were also supported by Algeria, Kuwait, Libya, Morocco, Pakistan, P.L.O, Sudan, and Tunisia. These countries gave money, weapons, and support to the Arab troops, so much against Israel, but after this, Israel gained territory.

In *Revelation 11 9-10*[9] **And for three and a half days, all peoples, tribes, languages, and nations will stare at their bodies. No one will be allowed to bury them**. I think there are some important lines there. All the tribes' languages and nations stare at the body. How can that be without having the invention of TVs. Let's remember that Revelation was talking about the end times. This is the last chapter of the Bible.

Daniel is one of the major prophets of the Bible. He is the one who prophesied Jesus' birth and death. Daniel can be argued for top prophetized. In *Daniel 12:4* **But thou, O Daniel, shut up the words, and seal the book, even to the time of the end: many shall RUN TO AND FRO, and knowledge shall be increased.**

Run to and Fro means long distances. Certainly, the internet has increased our knowledge. Daniele was talking about the end times.

In *2 Timothy 3:1-5* **This know also, that in the last days perilous times shall come. For men shall be lovers of their own SELVES, covetous, boasters, proud, blasphemers, disobedient to parents, unthankful, unholy, Without natural affection, trucebreakers, false accusers, incontinent, fierce, despisers of those that are good, Traitors, heady, high-minded, lovers of pleasures more than lovers of God.**

What could be more self-love than the selfie? In 2013, the word "selfie" entered the Oxford Dictionary. There has been a real increase in selfies since 2010, when the iPhone with a front camera was released. ⅓ of the pictures are selfies. Facebook is full of people who are posting selfies. Facebook's price has increased from $ 18 in August 2012 to $ 168 in August 2017. To put that in perspective, Coke stock went from 39 in August 2017 to 45 in August 201.

People who use selfies and use social media tend to suffer from depression. Millennials who use social media the most tend to be more depressed. It is theorized that when comparing your pictures to yourself, you get depressed. If you see someone on the beach with a big smiles, you get down. Few people, when they take selfies, don't have a smile. So when you're on social media, you think everyone is happy all the time. Social media is a facade.

In the same text, disobedient to parents, unthankful, unholy." There has been an increase in kids not minding their parents. The first case of a child emancipating their parents was in 1980. What other cases show more than not minding their parents? It's not uncommon to see talk shows with kids not minding their parents. With Maury episodes like My 14-Year-Old Daughter Claims She's a Pimp, Send Her to Jail! (November season 19 episode 37) or Help Us... Our Teens Are Over-Sexed and Out of Control (August season 19 episode 150) or Will Jail Tame My Wild Teen Daughter? (season 18 episode 82). There have been 19 years of episodes like those pictured above. These kids are demonstrating disobedience to their parents, unthankful.

As for unholy there are reports about less participation in church. Since 1995, over 1,000 Catholic churches have closed. There are many polls that show faith has gone down, There are more and more atheist congressmen, such as Barney Frank (D) of Massachusetts.

Another sign of the times in *2 Timothy 3-5 Message* **You're going to find that there will be times when people will have no stomach for solid teaching, but will fill up on spiritual junk food—catchy opinions that tickle their fancy. They'll turn their backs on truth and chase mirages. But *you*—keep your eye on what you're doing; accept the hard times along with the good; keep the Message alive; do a thorough job as God's servant.**

Doesn't that sound like today's time? People are telling more lies. Lies that fit their reality. How many marriages face infidelity where they lie. There has never been a higher divorce rate. There is so much lying that there are several TV shows about it. There is a TV show on MTV called Catfish. Catfish is a show about people who think their romantic partners are lying. Most of the time, it is a lie. They lie about their weight, size, gender, and relationship status. There is a current movement where they think the world is flat, and we are told lies about the earth being round.

In the Bible, it says *Revelation 13 15-17*[15] **And he had power to give life unto the image of the beast, that the image of the beast should both speak, and cause that as many as would not worship the image of the beast should be killed.**[16] **And he causeth all, both small and great, rich and poor, free and bond, to receive a mark in their right hand, or in their foreheads:**[17] **And that no man might buy or sell, save he that had the mark, of the name of the beast, or the number of his name.**

There has been an increase in radical religion, where if you are not a supporter you would be killed. Both Christianity and Islamic sects.

Jim Jones, a Christian, started preaching in the 1950's. Jim Jones left the United States after allegations of financial fraud and mistreatment of children. They moved to Jonestown, Guyana, in Africa. They lived in a Utopian society. There were some reporters, including a congressman who went there to investigate. Jim Jones felt the pressure and tried to kill the congressman. The report did a mass suicide where over 200 children were killed, and over 900 people were killed in total.

On September 11, 2001, there were over 2,700 people who died when 19 hijackers took over American Airlines flight 11 and United Airlines flight 175. The hijackers were masterminded by Osama bin Laden. A charismatic leader who wanted to kill Americans, whom he felt were unbelievers.

And he causeth all, both small and great, rich and poor, free and bond, to receive a mark in their right hand, or in their foreheads:[17] And that no man might buy or sell, save he that had the mark, or the name of the beast, or the number of his name.

Nobody could have imagined this many years ago. That people would not be able to buy or sell, but we have a society that is cashless. In Sweden, about 6/10 transactions were not using cash. In Sweden, cash circulation has been declining each year. With financial sabotage is it possible for people to stop people from buying and selling.

There have been improvements in technology, with implants available that are installed in people's hands. This is the mark of the beast. Some companies and businesses have created RFD chips that contain medical information. And there are companies that have implanted RFD chips implanted in the skin. Even our passports have chips in them.

This is what Jesus said about when back or when it is the end of days In Matthew 24 1- 25 **Jesus left the temple area and was going on His way when His disciples came up to Him to call His attention to the [magnificent and massive] buildings of the temple. [2] And He said to them, "Do you see all these things? I assure you and most solemnly say to you, not one stone here will be left on another, which will not be torn down."[3] While Jesus was seated on the Mount of Olives, the disciples came to Him privately, and said, "Tell us, when will this [destruction of the temple] take place, and what will be the sign of Your coming, and of the end (completion, consummation) of the age?"While Jesus was seated on the Mount of Olives, the disciples came to Him privately, and said, "Tell us, when will this [destruction of the temple] take place, and what will be the sign of Your coming, and of the end (completion, consummation) of the age?"[4] Jesus answered, "Be careful that no one misleads you [deceiving you and leading you into error]. [5] For many will *come* in My name [misusing it, and appropriating the strength of the name which belongs to Me], saying, 'I am the Christ (the Messiah, the Anointed),' and they will mislead many. 6 You will *continually* hear of wars and rumors of wars. See that you are not frightened, for *those things* must take place, but that is not yet the end [of the age].**

From the last question, the state of our world. We looked at wars. There are over 180 nations out of 190 nations that are in some kind of war. In the United States, there are about 7 major conflicts that the United States is currently in. We also talked about an increase in the number of people who proclaim they are God or God messenger.

Matthew 24 continues [7] **For nation will rise against nation, and
kingdom against kingdom, and there will be famines and
earthquakes in various places. [8] But all these things are *merely* the
beginning of birth pangs [of the intolerable anguish and the time of
unprecedented trouble].[9] "Then they will hand you over to [endure]
tribulation, and will put you to death, and you will be hated by all
nations because of My name. [10] At that time many will be offended
and repelled [by their association with Me] *and* will fall away [from
the One whom they should trust] and will betray one another
[handing over believers to their persecutors] and will hate one
another.**

There was also a line saying that in the last days there would be an increase in earthquakes; they described it as birth pains. There have been more intense earthquakes than there have been 3 8.7earthquakes or more in the last 20 years. Before that, 3 others in the last 370 years. With people living closer and in bigger populations, there will be more deaths because of earthquakes. Most recently, there was 200,000 people who died in the Haitian earthquake of 2010

Matthew 24 continues [11] **Many false prophets will appear and
mislead many.[12] Because lawlessness is increased, the love of most
people will grow cold. [13] But the one who endures *and* bears up
[under suffering] to the end will be saved. [14] This good news of the
kingdom [the gospel] will be preached throughout the whole world
as a testimony to all the nations, and then the end [of the age] will
come.[15] "So when you see the abomination of desolation [the
appalling sacrilege that astonishes and makes desolate], spoken of by
the prophet Daniel, standing in the Holy Place (let the reader
understand), [16] then let those who are in Judea flee to the mountains
[for refuge]. [17] Whoever is on the housetop must not go down to get**

the things that are in his house [because there will not be enough time]. [18] Whoever is in the field must not turn back to get his coat.

As I mentioned before, there has been an increase in the number of people who claim to be the messiah. Or they are God in human form. Christianity has been preached in all four corners of the world. Christianity has been a testimony to other countries.

Jesus was saying that when the end comes, it will be quick. There will be no warning, people will not have enough time to get their coat. The end of the world will be quick. People will have time to get away.

In the Bible *Zechariah 14 12-13* **[12] And the Lord will send a plague on all the nations that fought against Jerusalem. Their people will become like walking corpses, their flesh rotting away. Their eyes will rot in their sockets, and their tongues will rot in their mouths. [13] On that day, they will be terrified, stricken by the Lord with great panic. They will fight their neighbors hand to hand.**

In the Bible *Revelation 6 15 amp* **Then the kings of the earth and the great men and the military commanders and the wealthy and the strong and everyone, [whether] slave or free, hid themselves in the caves and among the rocks of the mountains.**

These references seem to suggest that there would be a nuclear war. A nuclear war would fit the story about rotting flesh and possible attack of muslim countries if somebody comes against Jerusalem. Later in Revelation, great men and military commanders will try to hide in the rocks of the mountains.

Conclusion

There have been major prophets throughout the Bible. The profits have been telling what has been happening ahead of time. Sometimes, to warn us, if we keep this behavior up, this will happen. Sometimes it's because he wants to tell us good things that are ahead of us. God tells us these things as proof as to confirm that he is God.

God

This is what Jesus said about when back or when it is the end of days In Matthew 24 1- 25 **Jesus left the temple area and was going on His way when His disciples came up to Him to call His attention to the [magnificent and massive] buildings of the temple. 2 And He said to them, "Do you see all these things? I assure you and most solemnly say to you, not one stone here will be left on another, which will not be torn down."3 While Jesus was seated on the Mount of Olives, the disciples came to Him privately, and said, "Tell us, when will this [destruction of the temple] take place, and what will be the sign of Your coming, and of the end (completion, consummation) of the age?"While Jesus was seated on the Mount of Olives, the disciples came to Him privately, and said, "Tell us, when will this [destruction of the temple] take place, and what will be the sign of Your coming, and of the end (completion, consummation) of the age?"4 Jesus answered, "Be careful that no one misleads you [deceiving you and leading you into error]. 5 For many will *come* in My name [misusing it, and appropriating the strength of the name which belongs to Me], saying, 'I am the Christ (the Messiah, the Anointed),' and they will mislead many. 6 You will *continually* hear of wars and rumors of wars. See that you are not frightened, for *those things* must take place, but that is not yet the end [of the age].**

There is more scripture than that. I just wanted to highlight some of the Bible verses we talked about in the last chapter. There are even more verses than what I described in this book. By some accounts, there are more verses about the end times than Jesus Christ. There are said to be over 150 references to the end of day reference.

FINAL THOUGHTS

_We started off with Is there a God. We proved there is a God. Mathematically, it is impossible for there not to be a God. We also took a look at evolution vs creation. We discovered the failures of evolution to come out with how we evolved.

The second question we looked at was the Bible: real and the scientific reasons why the Bible is true. The Bible had some scientific facts that we didn't understand until thousands of years later, such as washing hands with running water, quarantining from sick patience, and that blood is life. Without blood, we could not live. This seems common sense, but these are things that were found out thousands of years later, after the Bible was written.

We also explored the criticisms against the Bibles. Some of the critics of the Bibles pointed to the inconsistencies with the Bible. We pointed out that, compared with the New York Times, it really wasn't that inconsistent.

We then explored Jesus Christ biblically and non biblically. We separated into two questions because there was so much information. Biblically, we looked at the Old Testament, which showed how Jesus Christ was in the Old Testament. The Old Testament was carbon dated before Jesus Christ was born. In the Old Testament, it is stated that his hands and feet would be pierced, that he would be betrayed, and that he would be born in Bethlehem. Daniel even predicted when Jesus around the year that Jesus would be born.

We also looked at non-biblical sources such as Josephus, which was anti-Jewish. They not only talked about Jesus, but they also talked about other things in the Bible. Josephus also gave us details which are listed in the Bible such as John the baptist getting killed by Herod, and also it talked about Jesus Christ having a younger brother, James.

We also looked at why some people don't believe in God. Such as why God would allow suffering? We also answer why God would allow innocent people to do so. We don't see miracles, or why don't I see the wonders of God? Why don't my prayers get answered? We also looked at the trinity, which could be hard for even a senior Christian.

This book also looks at how to improve your life. It explored how to have a good Christian marriage and what are some things that cause divorce. This book also looked at what makes a good Christian and how to be a better Christian.

Some people don't believe in hell. If we believe the Bible and we believe in Jesus Christ. Then we must believe in hell. We must also believe in Satan. In the book, we also explained why Satan does not like humans.

In the last part of the book, we looked at how bad this world is. We looked at slavery, mass shootings, Income equality, running out of water, oil and gas, Phosphorus running out, Gold and silver, Natural disasters, animals disappearing, Wars, and abortions.

In the next chapter, we look at the world ending. They say that there are more prophecies about the day of reckoning (the end of the world). In the Bible they talked about the last days. In the last days, they talked about Israel being born in day, the Bible talks about coming from long distances on airplanes back to Israel. It also talked about televisions in the last days, the internet in the last days, selfies, and a generation that was very selfish. So these show that the end of the world is soon.

So if we know God is real, Jesus is real, and the Bible is true, we know that hell is real. We know Satan is real and hates human beings. We know the world is in a terrible place. We know the world is ending because the Bible describes modern life in televisions and planes.

If you remember the question how to become a better Christian, you must follow that :

1. Acknowledged that you sinned
2. Surrender to God
3. Love God with all your heart
4. Read the Bible and follow what it says
5. Get trained in the things of God (Church, Bible, videos, and podcast)
6. Allowing God to change them
7. Be accountable for your actions
8. Pray on a daily basis
9. Spread the Gospel

We are running out of time. We need to prepare for the end of the world. Joel 2:28"**And afterward, I will pour out my Spirit on all people. Your sons and daughters will prophesy,**

your old men will dream dreams, your young men will see visions.[29] **Even on my servants, both men and women, I will pour out my Spirit in those days.** [30] **I will show wonders in the heavens and on the earth, blood and fire and billows of smoke.** [31] **The sun will be turned to darkness and the moon to blood before the coming of the Lord.**

So at the end of the world, there will be Great things that God will show us. In the last days, there will be great miracles that God will show us. We want to be the sons and daughters who will perform miracles like the Bible says. We also don't want to be in hell. In the last days, there will be a divide between Good and evil. This will be the time you will decide on burning in a lake of fire or heaven. You could no longer say there is no God, no hell, no devil, no Jesus. You must decide between eternal life and death.

www.ingramcontent.com/pod-product-compliance
Lightning Source LLC
LaVergne TN
LVHW011519200926
848484LV00001B/75
* 9 7 9 8 9 0 2 3 5 1 9 8 6 *